Writing Technology in Meiji Japan

HARVARD EAST ASIAN MONOGRAPHS 387

Writing Technology in Meiji Japan

*A Media History of Modern Japanese Literature
and Visual Culture*

Seth Jacobowitz

Published by the Harvard University Asia Center
Distributed by Harvard University Press
Cambridge (Massachusetts) and London 2015

Printed in the United States of America

The Harvard University Asia Center publishes a monograph series and, in coordination with the Fairbank Center for Chinese Studies, the Korea Institute, the Reischauer Institute of Japanese Studies, and other faculties and institutes, administers research projects designed to further scholarly understanding of China, Japan, Vietnam, Korea, and other Asian countries. The Center also sponsors projects addressing multidisciplinary and regional issues in Asia.

This book was published with the assistance of the Frederick W. Hilles Publication Fund of Yale University.

Library of Congress Cataloging-in-Publication Data

Jacobowitz, Seth.
  Writing technology in Meiji Japan : a media history / Seth Jacobowitz.
    pages cm. — (Harvard East Asian monographs ; 387)
  Includes bibliographical references and index.
  Summary: "Boldly rethinks the origins of modern Japanese language, literature, and visual culture from the perspective of media history. This book represents the first systematic study of the ways in which media and inscriptive technologies available in Japan at its threshold of modernization in the late 19th to early 20th century shaped and brought into being modern Japanese literature"— Provided by publisher.
  ISBN 978-0-674-08841-2 (hardcover : alk. paper) 1. Japanese literature—1868—History and criticism. 2. Literature and technology—Japan—History—19th century. 3. Literature and technology—Japan—History—20th century. 4. Literature and society—Japan—History—19th century. 5. Literature and society—Japan—History—20th century. 6. Popular culture—Japan—History—19th century. 7. Popular culture—Japan—History—20th century. 8. Mass media and culture—Japan—History—19th century. 9. Mass media and culture—Japan—History—20th century. I. Title.
  PL726.6.J33 2015
  302.230952—dc23

                                                                          2015005307

Index by Mary Mortensen

Printed on acid-free paper

Last figure below indicates year of this printing
25 24 23 22 21 20 19 18 17 16

Contents

Figures

Acknowledgments

I first conceived this project while visiting Tokyo in the summer of 1999 on a pre-dissertation research grant from the Einaudi Center at Cornell University. I was casting about for new ideas after having just completed my Master's thesis on postmodernism in contemporary Japanese literature. Yano Yutaka, then an editor at Shinchōsha, introduced me to Friedrich Kittler's *Gramophone Film Typewriter*, which by coincidence had been translated that year into both English and Japanese. Kittler (1943–2011) refreshingly provided a methodology that not only synthesized ideas from many of the poststructuralist thinkers I was then avidly reading, but also helped make coherent a media history that I realized was similarly waiting to be recovered from Meiji era (1868–1912) Japanese archives.

It goes without saying that a very long list of individuals and institutions helped me along the way to completing this book. Prior to embarking on graduate work, I spent a year at Nagoya University on a Fulbright Fellowship reading Meiji literature with Wakui Takashi and Tsuboi Hideto. Professor Tsuboi continued to provide encouragement over the years, for which I am truly grateful. At Cornell University, I had the privilege of studying with four outstanding scholars: Brett de Bary in modern Japanese literature, Naoki Sakai in Japanese thought, John Whitman in Japanese linguistics, and Timothy Murray in English and film studies. I must add a fifth name as well: Frederick Kotas, former bibliographer of the Japanese collection at Cornell, who played a crucial role in bringing the Maeda Ai Collection to Cornell University. Fred was a constant resource and friend when I, like the hapless Sanshirō in Natsume Sōseki's novel of the same name, went in a perpetual search of books never before read or scribbled upon. I was also extremely fortunate to receive the support of a Japan Foundation Fellowship to Waseda University where I conducted research from October 2000 to May 2002 under the direction of Professor Takahashi Toshio. His guidance and unfettered access to Waseda's Meiji-era collection facilitated many of the archival discoveries discussed in this book.

In 2006–07 I was a postdoctoral fellow at the Reischauer Institute of Japanese Studies at Harvard University, where I gained valuable opportunities to present and revise portions of this book. A Northeast Asia Council grant from the Association of Asian Studies in the summer of 2009 and a Griswold Grant from Yale University in the summer of 2014 helped me to gather some final materials and insights which aided in its completion. This book was published with the assistance of the Frederick W. Hilles Publication Fund of Yale University.

I would be remiss not to thank my colleagues in the East Asian Languages and Literatures Department at Yale: John Treat, Edward Kamens, Aaron Gerow, Jing Tsu, Will Fleming, Kang-I Chang, Tina Lu, and Mick Hunter. I would also like to thank my former colleagues in the interdisciplinary Humanities Department at San Francisco State University, especially Saul Steier, Mary Scott, and the indefatigable George Leonard for their support during the lean years. Special thanks also go to Haruko Nakamura and the East Asian Library at Yale for helping me with the cover art and obtaining permissions for works in the collection.

The list of colleagues, mentors, and friends whom I wish to thank includes Miyako Inoue, Tatsumi Takayuki, Tim Screech, Michael Bourdaghs, Rebecca Copeland, Nakagawa Shigemi, Ted Mack, Alan Tansman, Tomotsune Tsutomu, Micah Auerback, Bruce Baird, Aaron W. Moore, Keith Vincent, Kuge Shu, Jon Abel, John Mertz, Jenine Heaton, Shiho Yoshioka, Dan McKee, Tom LaMarre, Alisa Freedman, James Dorsey, and Dennis Washburn. A tremendous debt of gratitude is also owed to my parents, Lawrence and Wilma Jacobowitz, and to my family for their support over the years. Without their help none of this would have been possible.

INTRODUCTION

Balloon Ride

The year is 1861 and technology is in the air. Two bearded Westerners in frock coats stand before a hillside, gazing toward a hot air balloon in which a third figure stands aloft holding two flags. A white, two-story building lies directly below the balloon, its lower story almost swallowed up by the slope of the hill. A scattering of clouds in the sky above and leafy treetops in the distance complete the picture. I begin with this scene, which appears in a woodblock print by Baisotei Gengyo (figure 0.1),[1] less for the spectacle of the flying machine it presented to the Japanese populace, than for the interconnected verbal, visual, and print regimes it represents during the interval between the American mission to Japan (1853–1854) and the dawn of the Meiji period (1868–1912). It sets the stage for this book to undertake a media history of modern Japanese literature and visual culture.

That the picture appears to float is due not only to the balloon, but also to the perspectival techniques of "floating pictures" (*uki-e*) that came into vogue in the woodblock prints of the late Edo period, derived in part from the illusion of perspectival depth in Dutch copperplate engraving.

---

1. Baisotei Genkyo (1817–1880) was best known as a book designer who collaborated with writer Kanagaki Robun. He also designed the title page of Hiroshige's *100 Famous Views of Edo* after the artist's death.

0.1 Baisotei Gengyo, *Shashinkyō fūsenzu* (Camera: A hot air balloon picture).
Photograph © 2015 Philadelphia Museum of Art.

There are further indications of Baisotei's desire to layer the print with multiple signifiers of Western art and writing technology. As Julia Meech-Pekarik explains, "to enhance the foreign scene, the artist provided a frame in imitation of an oil painting; some versions even have wax rubbed onto the surface in imitation of an oil painting."[2] On that ornate, floral border of the print is also the title *Shashinkyō fūsenzu* 写真鏡風船図 (Camera: A hot air balloon picture). This instance of bricolage situates the print squarely in an epistemic shift that occurred from the 1850s to 1870s. Many of the artists who pioneered the use of Western styles of oil painting and draftmanship also comprised the first generation to experiment with a host of intermediary genres that emerged from late Edo to early Meiji. They mixed and matched elements of woodblock prints, oil painting, lithography, typography, painting, and photography.[3]

Previously utilized to express the idealized essence of things rather than to capture the actuality of their external appearance, the concept of *shashin* (photography) and cognates such as *shasei* (sketching) and *shajitsu* (realism) would emerge within a new field of media concepts and practices that also radically redefined linguistic and literary discourse. It was part of a new signifying constellation that coalesced during the Meiji period and centered on *utsushi*, a term that contains the multiple meanings to write, copy, trace, inscribe, and project.

By the early 1860s when Japanese woodblock prints were picked up by the Impressionists in France and England, Europeans had already begun to create Japoniste styles in imitation of floating world prints (*ukiyo-e*), including the joyful mangling and spurious invention of Chinese characters as an atmospheric effect. In Japan, meanwhile, where access to images of European and American modernity were gleaned only from a small store of paintings, prints, and illustrated books, alphabetic writing could still be rendered as gibberish so long as few were literate in the strange, sideways written script. Baisotei was thus in good company when he printed the name of the balloon in capital letters across its midsection

2. Meech-Pekarik, *The World of the Meiji Print*, 20.

3. See Kinoshita's *Shashingaron* (On photographic-paintings) on the diversity of intermediary genres such as painted photographs, silhouette and shadow drawings (*kage-e*), and so-called "camera" kabuki actor prints (*haiyū shashinkyō*), which despite their ostentatious reference to photography, in fact combined elements of Dutch glass-plate and Japanese woodblock print styles.

as "CNИTCTUTION" instead of "CONSTITUTION." The disordering of Roman letters and inclusion of what appears to be a Cyrillic И were simply mistakes of his incomplete alphabetization. His print is based on an illustration in the travelogue *Futayo gatari* 二夜語 (Two nights' tale, 1861) by Katō Somō, a member of Japan's first embassy to the United States. As Meech-Pekarik reveals,

> On June 14, 1860 the Japanese embassy to the United States had paid a visit to the Philadelphia Gas Works to witness the ascent of two large balloons; the Constitution was piloted by Professor T. S. C. Lowe of New York and the other by Professor William Paulin of Philadelphia. . . . Ballooning had been popular in the West since the experiments of Montgolfier brothers in 1782, but the first balloon voyages in Japan did not take place until 1872, when three navy balloons were sent up in trial flight at Tsukiji in Tokyo.[4]

Comparison to the original illustration, which also clarifies the mystery of the two indeterminate flags as American, leaves no doubt as to the provenance of Baisotei's model.

Ultimately, the specifics of time and place were of lesser importance so long as a broadly Western or "Dutch" sensibility was imparted. In the lower right-hand side of the frame, in a third authenticating mark is the artist's signature, Baisotei *geboku*, or Baisotei's playful ink. This connects it to the episteme of the Tokugawa floating world in which amateurs held up exotic curiosities to the reading and viewing public. Baisotei's lettering, moreover, reveals a naïve fascination with the opacity of foreign scripts. The misspelled Constitution hovering in the air—a term doubly signifying the fundamental laws and principles by which a nation-state establishes itself, and the written record of said laws and principles—effectively underscores a political watershed that had yet to be enacted on Japanese soil. The result instead, apropos of the Tokugawa floating world's playful brush with Western modernity, is the floating signifier of nonsense letters on a floating balloon in a floating picture. Nor is it the only instance of writing scrambled and set adrift. The man's coat incorporates into its floral pattern a motif of letters that includes another *И* and *B*, an upside-down *A* and *T*, and a backward *C* in the

4. Meech-Pekarik, *The World of the Meiji Print*, 20.

center. Another row of letters run along the hem: a backward *D*, an *E*, an *L*, an upside-down *P* and *V*, and a final, indistinct letter. Their phonetic value is beyond the point, however, as they are purely decorative.

How quickly things would change in the years to come. A new investment in the transparency of phonetic scripts arose in the Meiji period as language reformers sought to mend the fragmented polity of Tokugawa Japan into a single, cohesive people (*kokumin*) united by a common national language (*kokugo*) and script (*kokuji*). The latter category applied not only to kana and roman letters, but also to experimental scripts, most importantly the shorthand notation (*sokki*) adapted from Western models and used in an effort to seamlessly record and transmit between spoken and written Japanese. A series of simple lines and loops that mapped onto phonetic values and could be written efficiently and accurately, shorthand was the consummate manual transcription system of the nineteenth century. It was no coincidence that in English it was dubbed "phonography" by its inventor Isaac Pitman, some thirty years prior to the invention of Edison's phonograph, or that in mid-1880s Japan, it earned the moniker "verbal photography" (*kotoba no shashinhō*). In a fitting resonance with the constellation of *utsushi*, practitioners of shorthand coined the catchphrase "to write things down just as they are" (*ari no mama ni utsushitoru*) to describe its high fidelity recording.

Initially introduced for the civic purpose of recording political speeches and the conduct of the state, shorthand was to garner more public attention when it was applied in the mid-1880s to the transcription of *rakugo* and *kōdan*, two types of popular theatrical storytelling. It was in this capacity that the serialized, shorthand transcribed pamphlets for Sanyūtei Enchō's *Kaidan botan dōrō* 怪談牡丹燈籠 (Ghost story of the peony lantern, 1884) enabled for the first time a form of unvarnished vernacular writing, the forerunner to what became known as *genbun itchi*, meaning literally, "the unification of speech and writing," or as I will refer to it hereafter, "the unified style." By dint of its origin in new forms of verbal and visual media capture, this literary mode was imbued with new qualities of indexical mimeticism. As I further argue in this book, it should accordingly be called "transcriptive realism" in recognition of its link to shorthand and the constellation of *utsushi*.

The genealogies of modern Japanese literature that commence with Tsubouchi Shōyō's literary theories in *Shōsetsu shinzui* 小説神髄 (Essence

of the novel, 1885) and Futabatei Shimei's "first modern Japanese novel,"
Ukigumo 浮雲 (Floating clouds, 1887), have long signaled *The Peony Lantern* as a source of fleeting inspiration, but have otherwise bypassed shorthand's compositional strategies and critical appraisal by Shōyō and others,
leaving it to haunt the margins of the canon as a ghostly remainder. While
"writing things down just as they are" would become the compositional
imperative of modern Japanese realism from the literary sketching (*shasei-bun*) movement initiated by Masaoka Shiki to modern novels by Kunikida
Doppo, Shimazaki Tōson, Tayama Katai, Natsume Sōseki, and many
others, its media-historical basis would be forgotten, lost, or marginalized
by successive generations of writers and scholars. Here, however, by building upon recent scholarship in media studies and media history, I seek to
reestablish the nascence of modern Japanese literature and visual culture
from within a field of techniques and technologies of writing.

Against the lingering ambiguity and play of signifiers in Baisotei's balloon ride, an unprecedented shift toward new standardizing measures of
time, space, and language, as well as a profusion of new media systems and
technologies, redrew the boundaries of perception and lived experience
from the 1870s to early 1900s. These developments had profound conse-
quences for inculcating national subjectivity and training a modern outlook onto the world. Thereafter, whether in literary or visual terms, a copy
would first and foremost be evaluated according to its indexical (calculable/
measurable) relation to external phenomena. As art historian Satō Dōshin
maintains, "For realism to truly become established, quintessentially
nineteenth-century factors were needed—a standard of values different
from idealization, beautification, or deification; a recognition of reality as
it is and the power to 'see'; and a new belief in one's visual abilities."[5]

We should not be misled into thinking that any form of media capture restores presence by representation, and language is no exception.
Miyako Inoue provides a brilliant précis of commonsensical notions of
language as a natural and innate conduit. According to such views,

Language is a transparent medium, purely and exclusively referential in its
function, according to which nothing comes between language and the
world; that is, there is an exclusive and context free, one-to-one correspon-

5. Satō, *Modern Japanese Art and the Meiji State*, 244.

dence between sound and the word, word and meaning, and language and the world. Language reflects what is already out there—always one step behind the world, docilely ratifying and confirming it. Such *a realist conception* of language is inherently ideological because it *effaces* the semiotic work of language in actively mediating and producing what is seemingly merely given, reversing the order of things as if the world existed *as it is* without the mediation of language. Linked up with the regime of modern power, language serves to turn things, categories, events, and ideas into a *fait accompli*.[6]

Effacement of the medium, or writing under erasure: although her words were not written with Meiji-era literary, linguistic and visual discourses in mind, they accurately describe a central belief that the invention of a new, transparently mimetic national language and script could serve as the basis for the realist novel and other contemporary genres. Memory of the medial processes would prove short-lived on any number of counts. By the early 1900s, the shorthand transcription of oral storytelling would recede from memory as younger writers simply embraced the unified style as a fait accompli in its own right. Eventually the question of national script would lose its equal status, too, and be subsumed into the field of national language. In other ways to be described later in this book, a literary canon was invented by future generations of scholars at the expense of these and other aspects of media history. It is therefore fitting that we begin at a point where matters of writing technology were still up in the air, prior to the moment when forgetfulness could set in.

Systems of Writing Things Down

This book investigates the discursive transformations that reshaped the literary, visual, and linguistic landscape of Meiji-era Japan (1868–1912). The critical approach of this book is principally indebted to the pioneering work of German media theorist and literary scholar Friedrich Kittler. Kittler adopts *Aufschreibesysteme* as the term of his methodology, which

6. Inoue, "Gender, Language and Modernity," 402 (my emphasis).

means "notation systems," or more literally, "systems of writing down." It was a neologism originally coined by the judge Dr. Schreber in his published memoirs, which were famously analyzed by Sigmund Freud. Kittler repurposes—or rather, restores to—the term its fullest signification, by using it to designate the proliferation of writing technologies and their effects in modernity.[7] Since the German term has no direct equivalent in English, Kittler's English translators have rendered it "discourse networks," and I make use of this critical, if not precisely synonymous term, at key junctures throughout my argument. As Kittler rightly observes, the discourse analysis developed by Michel Foucault takes as the limit of humanistic knowledge the primary medium of storage and retrieval available until the advent of modern media in the nineteenth century, namely the monopoly of print.[8] By contrast, discourse networks calls attention to material deployments of writing through channels of recording and transmission that are not limited to what Marshall McLuhan called the "Gutenberg Galaxy." Moreover, where discourse analysis implies objects that can be extracted from archives, as per Foucault's archaeology of knowledge, the concept of discourse networks offers a far more encompassing approach to identifying the often-fleeting traces of media.

This is not to say that Kittler's methodology is antagonistic to archival research or the representational codes that authors and texts lay bare. On the contrary, *Discourse Networks 1800/1900* sets up parallel instances of the discourses of Romanticism and Modernism in formation, whose constituent parts depend less on a particular genius for their creation (Goethe and Nietzsche, respectively), than a vast rearrangement of writing technologies, social relations, educational institutions, and the apparatus of the state. If I may take the liberty of borrowing from David Wellbery's impressive synopsis of the German episteme of 1800, he observes of the conventions for Romantic poetry, universal alphabetization, and other means by which national subjectivity was inculcated that

---

7.   In a 2006 interview with John Armitage, Kittler summarizes Schreber's psychotic worldview as follows: "By appealing to the notion of *Aufschreibesysteme*, the madman sought to imply that everything he did and said within the asylum was written down or recorded immediately and that there was nothing anyone could do to avoid it being written down, sometimes by good angels and occasionally by bad angels" (*Theory, Culture and Society*, 18).

8.   See Kittler, *Gramophone, Film, Typewriter*, 2–3.

they are discursive facts, nodal points in a positive and empirical discursive network, functions in a system of relays and commands that has no center or origin. As such they do not disguise a reality that is anterior to them and from which they would spring; they produce reality by linking bodies (e.g., the eyes and ears and hands of children), to the letter and to instances of power. Soon this system develops its own theory (a linguistics of the root and the verb), its own imaginary (Poetry as translation of the language of nature), its own protocols of reading (the Romantic hermeneutics of the signified). It realizes itself across institutional reforms (from primary schools to university lecture halls), it is codified in laws (the Universal Prussian Law of 1794 mandates both authorial copyright and maternal breast feeding), it shapes careers (as the new genre of the Bildungsroman reveals).[9]

National particularities aside, this model is not *structurally* very different from what transpired in late nineteenth-century Japan. Certain pathways and conduits that developed in the Meiji episteme closely track or run concurrently with developments in Germany and elsewhere in the West. Meiji Japan witnessed an array of national, imperial, and international standardization movements—temporal, spatial, and linguistic—in tandem with the new media technologies that increasingly redrew the boundaries of daily life across the globe. It likewise presided over the creation of a national postal service that regulated the sending and receiving of all written messages, including those conveyed via telegraph; national language and script reforms, including experimental phonetic scripts such as shorthand notation; and new categories of literary realism that culminate in the modern novel.

While Kittler's study has not yet been translated into Japanese, in critical writing about it, media studies scholar Tamura Kensuke chose to render the phrase "discourse networks" as *kakikomi shisutemu*, which literally means "systems of writing down." While this is an acceptable alternative, a more precise translation that does justice to the media-theoretical language rooted in the actual terminology of the Meiji era would be *utsushitori shisutemu*, which recuperates the catchphrase "to write things down just as they are." Indeed, the fact that this catchphrase insistently recurs across such a variety of texts and contexts is the mark of an episteme where writing, writ large, makes its presence felt everywhere.

9. Wellbery, "Foreword" to Kittler, *Discourse Networks 1800/1900*, xxiii–xxiv.

The title of this book, *Writing Technology in Meiji Japan*, is accordingly dialogic. It extends in the direction of both characterizing material conditions of writing and elucidating the self-reflexive, generative process of writing about technology in literature and visual culture. Working between these two levels, we see for the first time where the primacy of authors and texts, so often assured by canonical approaches to Japanese literary studies, takes a back seat to media as imbued with agency. Until recently successive generations of scholars were taught to read authors and texts in an almost purely exegetic capacity that gave little consideration to the vast enterprise of excavating media history. Against the grain of canonical genealogy, the media-historical origins of Japanese literary, linguistic and visual modernity not only demand their own archaeology; they also require that we resituate familiar authors and texts that have been thoroughly naturalized/nationalized as relays in media concepts, practices, and processes.

The Paper Trail

Media history has always had a larger intellectual foothold in Japanese-language scholarship of Japan studies than its North American counterparts. Kono Kensuke, Komori Yōichi, Kamei Hideo, Li Takanori, and many others have made invaluable contributions to historicizing the role of media concepts and practices in Japanese modernity. While I am deeply indebted to their respective analyses, which I have done my best to incorporate throughout this book, several differences distinguish my approach from previous scholarship. The early chapters in particular examine the standardization movements and new techniques and technologies of writing in Europe and North America that parallel, overlap, or in some cases are in direct dialog with the modernization and modernity of Meiji Japan. Without these distinctions in place the materiality of discourse networks fades into the background and can remain only at the level of metaphor in the pages of literary texts.

The early chapters likewise emphasize the extent to which Meiji Japan was in a close exchange of ideas and material culture with the Anglophone world. The incommensurability of speech and writing was not only a central problem for Meiji language reformers; it was a central feature of nineteenth-century Western discourse. In the Anglophone world, this gave rise to famous examples such as George Bernard Shaw's *Pygmalion* (1912), modeled after the irascible grammarian and philologist Henry Sweet (1845–1912). Remediating this gap was likewise the impetus behind experimental phonetic scripts such as Isaac Pitman's shorthand notation and Alexander Melville Bell's Visible Speech in the latter half of the nineteenth century. There is an all-too-facile assumption made in Japan studies that Japanese language reforms took place against a backdrop of fully formed or perfected Western languages. Nothing could be farther from the truth. From experimental phonetic scripts to the notion of a unified, vernacular language itself, Meiji reformers were in dialog with, not forever a step behind and struggling to "catch up" to, their Western, principally Anglophone, counterparts.

Last but not least, media history assumes a broader interrogative function than literary history alone. It was Foucault's radical insight into disciplinary formation that modern fields of knowledge are constituted from without. That is to say, they do not owe their origins to the putative objects they seek internally, but to external causes. Foucault's response was to attack this silence at the center that establishes myths of continuity and reveals its presence to itself. For proponents of the canonical accounts of the modern novel and the unified style as the basis of national language, their mutual blind spot is a question of scripts: namely, how phonetics were equated by Meiji intellectuals with the most advanced methods of media capture. This book in no way presumes the inevitability of the unified style or its role as the vehicle of realism in modern Japanese literature. Rather, it investigates a wide spectrum of debates and experiments that did not necessarily win the day, but nevertheless had lasting reverberations, and sometimes cascading effects, in the Meiji episteme.

With these qualifications in mind, it must be pointed out that the basic facts about the transcription of Sanyūtei Enchō's *The Peony Lantern* in Japan are well known and not, in and of themselves, controversial. Scholars and lay readers have long been aware that Tsubouchi Shōyō encouraged Futabatei Shimei to emulate Enchō's storytelling to achieve a modern style. The catch is that literary historians could always isolate this as an exclusive, one-time event—a strange occurrence, perhaps, but not a decisive turning point, much less an indicator that led to an

even more extensive groundwork of modernity. Still, even the basic prop-
osition turns out to be somewhat misleading. It was not Enchō's live performances that Futabatei had to witness in order to grasp his apparent literary
genius, but the shorthand transcriptions themselves. This provides an
important lesson about media that informs the narrative scope of this
book. I do not dispute that authors, innovators, and reformers can actively
participate and interact with one another in discourse. Rather, my focus is
upon the ways in which media mediate. The very existence of shorthand
transcriptions as an object to be imitated or adapted cannot be overstated.
In fact, shorthand transcriptions in unison with other systems of writing
down provided the means by which verbal, visual, and print regimes
could, for the first time, converge into "modern Japanese literature."

   This book attempts to provide a unified theoretical and archival
framework for understanding the media history of Meiji literature and
culture. The chapters form a logical sequence from the general medial conditions of the Meiji episteme (part I) to the centrality of experimental
phonetic scripts and the rethinking of conventional scripts in the debates
over national language and script reform (part II), to the emergence of
the unified style and transcriptive realism via phonetic shorthand (part III),
and their dissemination into modern Japanese literature (part IV). Collectively they demonstrate how a multiplicity of globally synchronic
media concepts, practices, and processes were assembled in Meiji. These
include standardization movements; the rise of new communications
systems such as telegraph and post; the aforementioned national language
and script reform; and new literary styles and modes of realism exempli-
fied by writers such as Masaoka Shiki and Natsume Sōseki.

   The first chapter examines the late nineteenth-century standardization movements that contributed not only to the consolidation of Japanese
national identity, but also to a new metrics of national-imperial time
and space. Starting with fundamental concepts of nationalism (Benedict
Anderson's) and technology (Martin Heidegger's), it periodicizes both
gradual changes and sudden ruptures that reshaped daily life and the disparate phenomena that had previously remained either intensely local or
unavailable to modern scientific methods. The second chapter amplifies
this discussion with a study of the discursive networks of telegraph and
post, which spread from the capital in Tokyo to the farthest peripheries
of the archipelago, the Japanese empire in East Asia, and beyond. It

focuses upon Maejima Hisoka, whose proposal to abolish Chinese characters sent to the shogunate authorities prior to becoming Japan's first postmaster-general is widely seen as the first salvo in the debates over national language and script reform. The chapter also evaluates Maejima's contributions to setting up the channels by which writing, print matter, and electronic data could be sent down the telegraph wires. It then surveys the encoded scripts and mechanical processes that recompose writing into a materiality prior to meaning, demonstrating the ways in which not only subjective agency, but the surveillance and control of the state are asserted through their channels. The third chapter provides a comparative analysis of the power of the written word to entrap within the boundaries of national space represented in Meiji by Mokuami's kabuki play *Shima Chidori Tsuki no Shiranami* 島千鳥月の白波 (Plovers of the island and white waves of the moon, 1881) over and against the conditions encapsulated in Hokusai's iconic woodblock print *Shunshū Ejiri* 駿州江尻 (Shunshū Station in Ejiri Province, c. 1830) from the collection *Thirty-Six Views of Mount Fuji*. At stake are the changing legal, social, and aesthetic codes put into the service of the state to regulate the conduct of its citizens.

The fourth chapter begins with Mori Arinori, for whom a national language and script based upon simplified English suggested the most expedient means of inculcating Japanese national identity and competing head-to-head with the Western powers. It observes how Mori's 1872 proposal was intimately bound up with spelling and phonetics reform in the Anglophone world. Directing his proposal in English to American scholars while serving as ambassador to the United States, Mori sought to overcome what he called the "hieroglyphic" nature of English spelling as well as the structural inadequacies of contemporary Japanese en route to establishing a modern state. In effect his proposal applied the same principle that comparable measures did with spatial and temporal units; in this case, its aim was to create a more perfectly standardized model of language. Despite the resistance and ridicule Mori endured by Anglophone and Japanese critics alike for suggesting to implement a consistent orthography of English, subsequent efforts to limit and regularize kana, including the elimination of variant forms kana, to limit the number of Chinese characters, and to create guidelines for romanized Japanese were consistent with the fundamentals of his proposal. The chapter concludes with the philosopher Nishi Amane, who insisted upon the materiality of writing, and specifically that of Roman letters, as the fundamental precondition for Japan's successful assimilation to the discourse of "Civilization and Enlightenment." Nishi correlated the production of modernity itself with the phonetic technology of the alphabet.

The fifth chapter explores Takusari Kōki's and his disciples' adaptation of Isaac Pitman's shorthand phonography (1882). Known as both phonography and a photographic method of words, shorthand contributed to a vast reorganization of economic, political, and literary activity by means of rapid manual recording and transmission prior to the popularization of mechanical audio-recording devices. It historicizes the origins of shorthand in the West, and devotes considerable attention to the competing theories espoused in Japan, particularly insofar as they point to shorthand's contributions to the debates over national language and script reform. A central premise of this book is that shorthand quite literally *underwrote* the unified style, whose ideal was to achieve the interchangeable states of "writing as one speaks" and "speaking as one writes."

The sixth chapter analyzes Alexander Melville Bell's *Visible Speech: The Science of Universal Alphabetics*, which was taught to Isawa Shūji directly by Alexander Graham Bell in 1878, and Isawa's subsequent adaptation and promulgation of Visible Speech. Isawa's applications of Visible Speech were integral to his implementation of national language education in the normal school system begun under Mori, including instruction for the deaf, disabled, and dialectically disadvantaged, and the earliest establishment of colonial education in Taiwan. Visible Speech for Isawa was ideally intended to regulate and advance the new concept of a unified Japanese national language (*kokugo*) at the center of an imperial system in East Asia. It was not Isawa's intention to create a hybrid or synthesis of colonial tongues or writing systems; rather, it was to maintain standardized modes of speech that could be taught to colonizer and colonized alike using Visible Speech. Japanese would thus become the hub by which each colonial language could maintain unilateral relations independent of one another. Although never put into practice in this fashion, Visible Speech is nonetheless revealing of the extent to which an imported, adapted phonetic script besides shorthand was advanced in the name of the Japanese imperial project.

The seventh chapter contextualizes the cognate constellations of concepts and practices of *utsushi* across painting, photography, and literature. It also carries out three close textual readings: in the visual regime, it focuses upon Ernest Fenollosa's polemical attacks in his presentation "*Bijutsu shinsetsu*" 美術真説 (The truth of art, 1882) against the word-image relations in literati art (*bunjinga*) en route to enforcing Western disciplinary boundaries. In the verbal regime it considers the accommodations made between *rakugo* and *kōdan* storytelling and shorthand notation evident in the prefaces, illustrations, and other framing devices of transcribed stories. Lastly, in the print regime it concludes with an analysis of Yano Ryūkei's political novel *Sēbe meishi keikoku bidan* 斉武名士経国 (Illustrious statesmen of Thebes, 1883–1884). This best-selling text, which was set in ancient Thebes and strove to represent the possibilities of political transformation espoused by the People's Rights Movement, was also a striking demonstration of the possibilities of shorthand for reconceptualizing the relations of political thought and literature. While Yano employed a mixed style based on classical Japanese grammar, he also enlisted the participation of two shorthand reporters to transcribe the two volumes of the text. In the first edition, the second reporter Wakabayashi Kanzō assisted Yano in providing the reading public with an afterword that demonstrated shorthand alongside kana-only and mixed kanji-kana scripts. In less than a year's time, Wakabayashi would also participate in the transcription of Enchō's *The Peony Lantern* and write a preface of his own alongside one by Harunoya Oboro, the pseudonym of writer and literary theorist Tsubouchi Shōyō.

The eighth chapter examines the three canonical texts typically placed at the forefront of modern Japanese literature: *The Peony Lantern*, *Essence of the Novel*, and *Floating Clouds*. Despite the rhetoric of a transparency bordering on hallucinogenic mimesis put forth by shorthand practitioners and early literary theorists, the status of the represented text vis-à-vis an original presence or site of enunciation was constantly fraught with the ghostly interference of the medium. This is not to replace one origins narrative with another in an endless regression toward ever earlier, but never quite definitive beginnings, but to reassess the archival and methodological underpinnings upon which the very concept of a canonical origin is secured. It is also to challenge the hermeneutic effects that proceed from the canonical trajectory: while many authors and texts retain their national-literary stature, the horizon of their legibility is irrevocably transformed.

The ninth chapter examines various writings from Masaoka Shiki, from his early iconoclastic attacks against conventional *haikai* poetics and reform of haiku and tanka to literary sketching in prose. I use the title from Shiki's early collection of essays, *Fude Makase* 筆まかせ (Scribblings, 1884–1892) as a point of departure and organizing trope that locates his work in the discursive ruptures and peripatetic movements that led to experiments with transcriptive realism in the unified style. Yet I also wish to account for Shiki's back-and-forth dialectics of life and literature, blood and ink. There is a conceit upon which scholars depend, the Romanticism of excavating the archive, which will yield from its dusty crypts every secret thought and feeling of its authors and texts as a consistent and coherent system of meaning: the text of life. Shiki plays into such a conceit as a paragon of Romanticism on several levels. His obsession with writing things down, a mania that we might diagnose as *Aufschreibesysteme*, was "inherited" by Naturalist writers in the late 1900s after his death, which recast physical and psychological phenomena in a scientific, or at least social Darwinian, perspective. There was also the matter of his tuberculosis, which manifested around 1895 and grew steadily worse, with a prolonged period of deterioration from 1898 to 1902. In spite, or perhaps because, of his physical debilitation, Shiki's experimentation with literary form and style continued unabated.

The final chapter explores Sōseki's *Wagahai wa neko de aru* 我輩は猫である (I am a cat, 1904–1906), which was at once the apotheosis of the unified style in the modern novel and its most compelling critique. Drawing upon the full complement of discourse networks at his disposal by the turn of the twentieth century—shorthand, literary and artistic sketching, other media technologies, and the production of humanistic knowledge in general—Sōseki calls attention to the limits of representation, recording, and transmission through the medium of a cat. With its feline amanuensis constantly disrupting the narrative flow with reflections upon the medium of writing itself, this text exemplifies the discourse of transcriptive realism whereby the modern self is always preceded, if not in fact constituted, by a scene of writing.

PART I

Discourse Networks of Meiji Japan

CHAPTER I

Standardizing Measures

Standing-Reserve

While the eighteenth and nineteenth centuries marked the expansion of European nation-states whose colonial and imperial ventures began the imposition of the cultural hegemony of the West on a global scale, the latter half of the nineteenth century witnessed unprecedented standardizations of time, space, and language within and across national boundaries. These constituted an effort not only to impose order on the contact zones where different cultures and peoples came together, but also to bridge the intensely local, yet fragmented, relations dispersed across the globe. Accordingly, Michael Adas, for one, has provided valuable contextualization for the ways in which universalistic, "Western" tropes of science and technology served to justify the colonization of Africa, South Asia, and China.[1] Apart from the ideological remarking of cultural difference, however, there has been far less acknowledgment paid in national or world histories to the many ways in which standardization distributes and defines the experience of modern life.

---

1. Adas, *Machines as the Measure of Men.* Although Adas mentions Japan, he provides only a cursory overview of favorable European attitudes toward Japanese industriousness and military prowess ranging from Francis Xavier in the late sixteenth century to Lafcadio Hearn at the turn of the twentieth.

Certainly standardization was to have profound implications for "the Great Japanese Empire" (*Dai nihon teikoku*), as the Meiji state was consecrated in the 1890 Constitution.[2] For the first several decades of the new regime, Japan struggled to repeal the unequal treaties that had been imposed upon the shogunate in the 1850s, even as it embarked upon modernization campaigns at home and on acquiring its own dominion in East Asia and the Pacific.[3] In order to properly evaluate the emergence of national language and modern literature in this era, it is necessary to situate them against the backdrop of an international milieu of standardizing measures and in relation to emerging media technologies. This chapter further seeks to identify how the rhythms of daily life in Meiji Japan were converted into the calculable units and scenes of writing of the industrial machine age.

Benedict Anderson's well-known definition of the nation as an imagined community depends upon a radical break from the time-consciousness of premodern societies, whose cosmologies harmonize the present with the past through the language of classical and sacred texts, to the horizontal fraternity of the modern nation-state that synchronizes and locates itself through vernacular print matter, notably the novel and newspaper. Anderson adduces to the temporal contrast between premodern and modern cosmologies the modern science of standardized measure:

> What has come to take the place of the medieval conception of simultaneity-along-time is, to borrow again from [Walter] Benjamin an idea of "homogeneous, empty time," in which simultaneity is, as it were, transverse, cross-time, marked not by prefiguring and fulfillment, but by temporal coincidence, and measured by clock and calendar. . . . Why this

---

2.   Modern use of *Dai-nihon teikoku* dates back at least to the early 1870s, as demonstrated by the fifth installment of Tsuda Mamichi's "On Government" in the *Meiroku zasshi* 明六雑誌 (1873). Tsuda affirms the basic right of ownership of land to private individuals by the benevolence of the emperor, asking the rhetorical question "Are there any among the people of the Great Japanese Empire who are not deeply moved by the broad and boundless grace of the emperor?" (Braisted, ed., *Meiroku Zasshi*, 200).

3.   Existing treaties with the major Western powers ended in the period 1894–1895 and were replaced over the ensuing years by the ratification of more equitable arrangements. Meanwhile Japan began its colonial expansion with the tributary states of the Ryūkyūs (Okinawa) and Ezo (Hokkaidō) in 1869, took over Taiwan after war with China in 1894–1895, and obtained substantial concessions in Korea and northeastern Asia after its narrow victory over Russia in 1904–1905.

transformation should be so important for the birth of the imagined community of the nation can best be seen if we consider the basic structure of two forms of imagining which first flowered in Europe in the eighteenth century: the novel and the newspaper. For these forms provided the technical means for "re-presenting" the kind of imagined community that is the nation.[4]

I will return later in this chapter to the question of how the clock and calendar were essential to the ordering of modern subjectivity. For their part, the newspaper and novel proved highly attractive to contemporary Western and Japanese literary scholars alike, who needed to look no further than the texts they were already reading for confirmation of how the nation came into existence. Yet Anderson's premise should be a starting point, rather than the last word, of a discussion of the media that structure national belonging—indeed, one would be remiss not to reread the above passage and emphasize the phrase *technical means*.

The modern state's efforts to convert premodern temporalities into a simultaneous production of newness *and* antiquity define the teleological double movement of national history. Yet the ratification of standard time (1884), of the metric system (1891), and of other international conventions of unitary measure in the late nineteenth century also began to radically recode what were once diffuse local practices and natural phenomena into a universal worldview. We need look no further than Martin Heidegger to furnish the underlying concept. In "The Question Concerning Technology" (1954), Heidegger insists that the essence of technology (*techne*) inheres not in any specific technology, but rather in its capacity to purposefully reveal and transform material, form, intent and ends into a systematic availability that he calls "standing-reserve" (*Bestand*). Heidegger argues for a manipulation of potentialities such that "everywhere everything is ordered to stand by, to be immediately on hand" for human use.[5] Yet, by extension, this process can also bring about relations of commensurability and exchange value where none previously existed. Through standing-reserve the inducement to conform to a common register (capitalism, nationalism, and so on) is thus brought to bear

4. Anderson, *Imagined Communities*, 24–25.
5. Krell, ed., *Martin Heidegger: Basic Writings*, 322.

upon nearly all aspects of human being. It is the logic of an increasingly homogeneous global system of quantification, commodification, circulation, and exchange.

Heidegger had already taken up principle of standardization in "The Age of the World Picture" (1938) in which he argued that a transformation based on exactitude of measure has transpired in the modern age: "When we use the word 'science' today, it means something essentially different from the doctrina and scientia of the Middle Ages, and also from the Greek *epistēmē*. Greek science was never exact, precisely because, in keeping with its essence, it could not be exact and did not need to be exact."[6] He emphasizes that what instead underlies the modern age is the specificity of mathematics and physics, for which measurement and calculation of numbers are the principal modes of determination. Moreover, although he is careful to differentiate science from machine technology, he makes a powerful statement about the new episteme the latter has brought about:

> Machine technology is itself an autonomous transformation of praxis, a type of transformation wherein praxis first demands the employment of mathematical physical science. Machine technology remains up to now the most visible outgrowth of the essence of modern technology, which is identical with the essence of modern metaphysics.[7]

To equate the metaphysics of the modern age with machine technology is thus to acknowledge a decisive break with the past. Perhaps more startling is the autonomy he accords machinery as coming to bear historical agency in its own right.

The principle of standing-reserve manifested through machine technology is revealed with remarkable consistency as a relation to writing. The suffix "-graphy" that is attached to all manner of instruments for recording and transmitting recapitulates "writing" in the broadest possible sense: collapsing distance with electric telegraphy (1835); capturing light with photography (1839) and moving images with cinema (1894); spreading the word via steam-driven rotary printing presses (1814) and

6. Ibid., 48.
7. Heidegger, "Age of the World Picture," in *The Question Concerning Technology and Other Essays*, 116.

typewriters (1870s); sound recording with the phonograph and transfer of the voice over the telephone (1877).[8] The impetus behind these and innumerable other media was to collectively reorganize disparate phenomena into analog(ous) forms of media capture.

Whether writing was done manually or by machine, one such outcome was the deployment of new scripts and codes introduced as prostheses to compensate for physical impairment or disability. Their dissemination from the nineteenth century onward precipitated various ways of recoordinating and standardizing relations between human and machines. Before Morse code (1836) or Pitman's phonographic shorthand (1837), the reading-writing system of Braille (1834) granted literacy to the visually impaired via a matrix of type pressed into raised dots and read by fingertip. By century's end, the trend turned toward making the visually and audibly impaired into operators of heavy machinery, which is to say, putting them to work as typists. Among the earliest models of the typewriter was Hansen's Writing Ball (1870), designed by the Danish principal of the Royal Institute for the Deaf in Copenhagen, Rasmus Malling-Hansen, so that the deaf and dumb might "speak with their fingers." Kittler reminds us that the Writing Ball was briefly but decisively put to use by the almost-blind Friedrich Nietzsche in 1882, who typed out on the cumbersome keys a simple statement of his experience interacting with the typewriter: "Our writing tools are also working on our thoughts."[9] Kittler finds in this famous statement a new set of limits for human subjectivity bounded by writing technology: "Writing in Nietzsche is no longer a natural extension of humans who bring forth their voice, soul,

---

8. In general, the dates given here refer to commonly attributed dates of invention, not of patent issuance or mass production. Of course, even this greatly oversimplifies the extent to which they were conceived as improvements or adaptations of existing technologies. Likewise, we should not forget the parallel or competing efforts to produce similar results by following common scientific principles. For instance, although technologies such as modern movable type and photography were introduced to the shogunate by the Perry Expedition in 1853, experimentation with daguerreotype technology was covertly funded by opposition leaders such as Shimazu Nariakira, the daimyo of Satsuma from 1851 to 1858.

9. Kittler, *Gramophone, Film, Typewriter*, 200. On the other hand, we might say Samuel Morse stole a march on Nietzsche in the delivery of telegraphic aphorisms by transmitting "What hath God wrought?" as the first official message sent by the U.S. government-funded telegraph in May 1844.

individuality through their handwriting. On the contrary: . . . humans change their position—they turn from the agency of writing to become an inscription surface."[10] This change in position portended more than a shift from abstract thinking to industrious, or industrial, typing. The bodily relationship to media, and media's workings on the body and mind, would be duly marked as that which occurs prior to any consideration of the writing subject as the author of textual meanings.

By the same measure, the ease and durability of media ensured conformity to new programmatic activities of writing for the disabled and able-bodied alike. Phonetic shorthand involved various strategies for positioning eyes, ears, and hands into linked circuits for maximum efficiency of recording. A more enduring example is the 1875 Remington typewriter's rearrangement of the alphabet into "QWERTY" which repositioned letters according to the statistical frequency of hit strokes, maximizing the productivity of ten fingers working in unison compared with the cluster of twos and threes when hand-writing by pen or brush. The standardization of the typewriter is also the standardization of the typist.[11]

Nietzsche had simply confirmed what had already been under way for nearly forty years, that is, the redeployment and safe return of phonetic or ideographic scripts into a new alphanumerical metrics. Shorthand notation resolves phonemes into a continuous stream of simple straight and curved lines, dots, and dashes: a record of material traces presumed, in advance of the phonograph, to permit an exact, if largely noiseless,[12]

10.   Ibid., 210.

11.   Equally important to the longevity of the typewriter was the spacing of the fingers to avoid repeated stress on the keyboard's internal mechanisms caused by certain clusters of frequently used letters in the alphabet. The distribution of those letters thereby diminished the likelihood a machine would break down and need to be repaired. It is only the center of the keyboard that preserves a cluster in alphabetic order, consisting of the consonants *d, f, g, h, j, k,* and *l.*

12.   The scratching of shorthand onto paper is a much different matter from the scratches that disrupt the smooth circular movements of a phonograph needle over a record. The recurrent conceit in the fidelity of shorthand recording is, of course, its ability to "capture things as they are" (*ari no mama ni utsushitoru*). Regardless of the extent to which shorthand exceeded earlier forms of transcription, it remained a highly precise mode of isolating sound from noise, sense from nonsense, and so on. In this respect, it could not approximate the mechanical effects of the phonograph and subsequent recording devices that captured the real as an unfiltered totality.

recording. Conversely, Morse code turned phonetic scripts into a code that might be transmitted down the wire as electric impulses by technicians who also "spoke with their fingers." Over the course of the nineteenth century, the deployment of writing as a sum of "–graphies" dominated immense areas of nation- and empire-building, even as it transformed the experience and expression of daily life. To truly appreciate these changes in context, I want next to investigate how the standards and conventions of time, space, and language were configured to become the operating systems in which machine technologies thrived.

Standards and Conventions

The first standardizations of spatial and temporal measure are often attributed to the rapid growth of the railroad industry in the early nineteenth century. Wolfgang Schivelbusch has famously described the "machine ensemble" of the railroad where the train as a mode of conveyance and its tracks as a route became inseparably associated as a single, indivisible unity. By extension, the interior of the train became a compartmentalized space furnished with panoramic views whose comings and goings create a vista onto national landscapes even as it was itself a prominent symbol of their industrial transformation.[13] Yet the machine ensemble of the railroad was not only a transportation network in the narrow sense; it was also inextricably linked to the burgeoning communications networks of telegraphy and signaling systems.

Although temporal and spatial measures remained intrinsically local means of coordinating knowledge and human experience throughout the nineteenth century, an epistemic rupture was registered by the intrusion of the railroad into the countryside in Western Europe and North America as early as the 1840s. As Michael O'Malley has observed, Henry David Thoreau, whose retreat to the woods was predicated on forsaking

---

13. One of the chief obstacles to the train creating international landscapes was the lack, or deliberate avoidance, of standardized measures for tracks and machine parts across borders. Consequently, a train could travel only as far as the tracks for which its wheels were fitted.

"mechanical aids,"[14] would observe from Walden the railroad's impact on rural life less in terms of a criticism of industry than for its capacity to replace fluctuating and seasonal patterns in nature with unwavering consistency: "The startings and arrivals of the cars are now the epochs of the village day. They come and go with such regularity and precision, and their whistles can be heard so far, that the farmers set their clocks by them, and thus one well-conducted institution regulates a whole country."[15] Hence, several decades prior to efforts to impose national or international standards on timekeeping by government and transportation industry officials, railroad networks restructured time-consciousness wherever they inscribed themselves into the fabric of daily life. While Thoreau generally approved of the punctuality applied in "railroad fashion" as conducive to spiritual and physical discipline, he decried the technologized image of the nation that was already being imposed on the national imagination as a falsely shared experience: "Men think that it is essential that the Nation have commerce, and export ice, and talk through a telegraph, and ride thirty miles an hour, whether *they* do or not."[16]

Given the centrality of the railroad to American industrial might and territorial expansion, it should not be altogether surprising that it was first introduced to Japan as a quarter-scale model presented to the shogunate by Commodore Perry's expedition in 1853. It was a demonstration of technological mastery that stood in stark contrast to the stagnation of Japanese innovation under the Tokugawa regime. This inertia would persist in spite of the toy train's dramatic appeal (some dignitaries from the shogunate insisted on straddling and riding it themselves), as no attempt was made until after the Meiji Restoration to create a Japanese rail infrastructure. It would be decades later still when standardizations of time and measure would be implemented.

---

14. Thoreau, *Walden*, 83. This resistance to machine technologies directly precedes the celebrated passage in which Thoreau declares, "I went to the woods because I wished to live deliberately, to front only the essential facts of life, and see if I could not learn what it had to teach, and not, when I came to die, discover that I had not lived" (83).

15. Ibid., 108. Also cited in O'Malley's richly anecdotal account of American railroads and standard time, *Keeping Watch*, 67.

16. Thoreau, *Walden*, 84 (emphasis in original).

The first line connecting Tokyo with nearby Yokohama was built in 1872 with infusions of British technical expertise and capital. Although strategic points across the main islands were rapidly connected by rail within the decade (with foreign assistance divvying up the geographic spoils), rail lines in Japan lagged considerably behind comparably industrializing nations well into the 1910s, a situation owing partly to the financial constraints from a succession of wars—the Satsuma Rebellion in 1877, the Sino-Japanese War in 1894–1895, and the Russo-Japanese War in 1904–1905—and partly to high engineering costs due to the geography of the mountainous archipelago.[17]

This is not to say that alternative modes of transportation and communication did not exist in the intervening years between the signing of unequal treaties in the 1850s and consolidation of centralized infrastructure in the 1870s. Prior to the completion of the Shinbashi-Yokohama line, enterprising foreign entrepreneurs began shuttling passengers and baggage by horse-drawn coach. The private British company, J. Sutherland & Co., established routes running from Odawara to Yokohama and Yokohama to Tokyo. Prior to the differentiation of transportation and communication networks that mark the beginnings of postal systems and discourse networks,[18] it was a matter of course that Sutherland & Co. bundled mail service into its carriage routes, carrying messages as well as persons and baggage. The first postage stamps in Japan were not, in fact, created by the Teishinshō, as the postal service was then known, but were the "1 and ¼ Boo" (*bu*) stamps printed by Sutherland & Co. in 1872.[19] Meanwhile, a carriage service established in the treaty port of Hyōgo (present-day Kobe) in 1868 by Rangan & Co. expanded its business to Tokyo in the following year under the name the Edo Mail. It earned the ire of Maejima Hisoka and other government officials for its reckless drivers, frequent traffic accidents, and aggrandizement of what should, in their view, properly have been the exclusive preserve of the Japanese government.[20]

---

17. Ericson, *The Sound of the Whistle*, 69–73.
18. Kittler and Griffin, "The City Is a Medium," 723.
19. Shinohara, *Meiji no yūbin, tetsudō-basha*, 6–7.
20. Ibid., 26–27.

One of the enterprising young Japanese who refused to cede this lucrative business to foreigners was Shimooka Renjō, who would later earn his fame as arguably the most accomplished early Meiji photographer. Along with seven partners, Shimooka founded the Narikomaya carriage service in 1869, which worked the Tokyo-Yokohama route until the railroad line opened. Shimooka funneled his profits into opening his own photography shop in Yokohama in 1873, where he taught photographic techniques to young artists, including Kobayashi Kiyochika, Takahashi Yuichi, and Yokoyama Matsusaburō. Shimooka's choice of location was hardly serendipitous. The capitalist and creative synergies afforded by new modes of transportation, communication, and media technologies stemmed from a ready source of materials and markets in the foreign settlements around the treaty ports. This included the Tokyo Tsukuji Type Foundry brought to the area from Dejima (Nagasaki) by Motoki Shōzō in 1872. Hence railroad, post, photographic, and movable-type printing facilities were densely concentrated in the transnational commerce of the Tokyo-Yokohama circuit, and many of the central figures associated with industrialization and modernization were in close proximity, if not, in fact, in mutually beneficial relationships.

These epistemic transformations were subsequently reflected in Meiji literature as well. By the late 1890s, when the railroad had already made considerable inroads into the countryside adjoining Tokyo, Thoreau's observations about nature and industrial society would be echoed by one of his avid young Japanese readers. Kunikida Doppo (1871–1908), whose prose fiction writing proceeded from experimentation with Romanticist free verse poetry to the literary sketching pioneered by Masaoka Shiki, was among the first to explore the new measures of space, time, and language in the modern realist novel. Published in the journal *Kokumin no Tomo* (The nation's friend) in 1899, Doppo's short story "Musashino" is narrated by a protagonist who sets out for the site of the eponymous ancient battlefield now covered by forests, guided by a map from the Edo period. He does so less out of an urge to pay homage to the layers of history associated with traditional *waka* and *haikai* poetics than from an attraction to the vestiges of the past and a landscape reclaimed by nature. He seeks to experience with his own eyes what cannot be grasped by older forms of representation. Nor is he alone in this sentiment, as he self-reflexively indicates: "The desire to see what remains of the Musashino we visualize through pictures and poems is not exclusive to me by any means."[21] True to the tenets of literary sketching, Doppo's narrator cannot rest content at the level of pure, immediate experience, but takes down his observations as entries in a diary, recording the transformations in the scenery according to time of day, changes in seasons, weather, and so forth, and thereby placing himself, as Karatani Kojin argues in his thesis on the discovery of landscape and subjective interiority, as much in the capacity of the scientist as the literary humanist.[22]

Musashino is not just outlying forests and fields for the narrator, but the ebb and flow of Tokyo, the rivers and footpaths crisscrossing between natural and man-made environs. Admonished by an old woman at a teahouse in the nearby village of Shibuya for coming out of the viewing season for cherry blossoms, the narrator is, of course, keenly aware that his fascination with a natural space reclaimed by forest marks a break with earlier cultural aesthetics. Yet this, too, will soon be superseded by industrial and urban encroachment. The sound of the steam whistle portends the eclipse of both the older way of life in Shibuya and the Romanticist idyll he discovers in the woods:

> In the summer when the nights are short and the dawn comes early, by sunrise the wagons are already beginning to pass by. All day long the rumbling wheels never cease. By nine or ten in the morning the cicadas have begun their chirruping from the high trees and it gradually grows hotter and hotter. The horses' hooves send up clouds of dust which are fanned into the empty skies by the wheels of the wagons, and flies flit from house to house, from horse to horse. Then one hears the distant boom of the noon gun, and somewhere from the skies over the city one hears the *blast of the steam whistle*. This is Musashino.[23]

In the span of several decades Shibuya would be absorbed into the ambit of the Tokyo rail systems, the forests giving way to the southwestern terminal hub of the imperial metropolis.

21. Doppo, *The River Mist and Other Stories*, 97. In Japanese, see *Kunikida Doppo Zenshū* (hereafter *KDZ*), 2:65.

22. Karatani, *The Origins of Modern Japanese Literature*, 22–44.

23. Doppo, *The River Mist and Other Stories*, 112 (my emphasis). I have modified Chibbett's translation by substituting the literal meaning of "blast of the steam whistle" for *kiteki no hibiki*, which Chibbett renders as "midday siren." See *KDZ*, 2:86.

The noon gun and the steam whistle mark time over and against the layered visualizations of the ancient site. It was already a decade and a half since the railroad networks in Europe and the United States imposed new time pressures upon communities long secure in their localized experience of natural time as the sun's rays moved across the earth's surface. However, as the scale of railroad travel increased, the contact zones of modernity exponentially multiplied into possibilities of very real collisions. And yet even taking for granted the safe passage of commuters and cargo, there was no limit to the disorder of rail traffic. Reacting to these pressures, standard time was in principle adopted in England in 1880, followed by the General Time Convention convened by an association of railroad companies in the United States in 1883, which implemented so-called "railroad standard time" across the *thirty-eight* contiguous states. Although England occupied a single time zone, there was a demand in the rapidly expanding United States to improve the efficiency of vast operations taking place over the expanse of a continent still fragmented by a dizzying array of local measurements. Standardization meant that time would no longer be defined by the gradual shift as one moved along the curvature of the earth, but as a series of geographically fixed time zones. As Clark Blaise observes, "For the first time in history, Boston and Buffalo, Washington and New York, Atlanta and Columbus, San Francisco and Spokane, all shared the same hour and minute. It didn't matter that Boston would be bright with the new day while Wheeling was still dark. In fact, it didn't matter what the sun proclaimed at all. 'Natural time' was dead."[24] Of course, railroad traffic alone does not explain the intensity that gathered around time-consciousness in the mid-1880s. In addition to the need for consolidating nautical and astronomical reference points for sea trade, the first transatlantic cables were laid in 1858 and 1866, with other international cables linking up much of the world overland or undersea in piecemeal fashion by century's end. The expansion of the postal system from local and regional levels to a truly global scale likewise contributed to the dynamics of standard, national time.

Building on these developments, the Prime Meridian Conference (also known as the International Meridian Conference) attended by

---

24. Blaise, *Time Lord*, 103. See also Schivelbusch, *The Railroad Journey*, 41–43.

delegations from twenty-five nations, including Japan, ratified world standard time in October 1884. With the transit instrument of the Greenwich Observatory as the principal reference point, it established the first universal day, ending reliance, or for lack of an alternative, dependence, upon solar time. Its equivalent in Japan was established in 1886 in the town of Akashi in Hyogo prefecture, which is conveniently located at 135 degrees east longitude and was subsequently given the moniker *toki no machi*, or "time town." The establishment of Japanese standard time in 1896 was used to set a western time zone for the outlying islands of the archipelago and newly colonized territory of Taiwan, and again expanded in 1910 to include Korea under Japanese Central Standard Time.[25]

From a strictly scientific standpoint, the assignment of a prime meridian over all others is at best an arbitrary human convenience, at worst a political imposition, as the French delegation and its supporters maintained in vain against the implementation of an English or "Anglo-American" standard.[26] Indeed, it was the mission of the anarchists popularized in Joseph Conrad's *The Secret Agent* (1907) who set out to blow up the Greenwich Observatory to strike a blow at the "sacrosanct fetish" of modern science it represents.[27]

Precise determination of a universal standard for temporal and spatial units was increasingly arrived at in the latter half of the nineteenth century through formal agreements implemented beyond the local and

25. Li, *Hyōshō kūkan no kindai*, 182.
26. Ibid., 199–209.
27. As the Russian *agent provocateur* Mr. Vladimir explains to the protagonist Verloc, an attack against science is the ultimate violence against modern civilization, and hence calculated for maximum impact beyond political or religious assassination, or overt statements of class struggle:

> Murder is always with us. It is almost an institution. The demonstration must be against learning—science. But not every science will do. The attack must have all the shocking senselessness of gratuitous blasphemy. Since bombs are your means of expression, it would be really telling if one could throw a bomb into pure mathematics. But that is impossible. I have been trying to educate you; I have expounded to you the higher philosophy of your usefulness, and suggested to you some serviceable arguments. The practical application of my teaching interests you mostly. But from the moment I have undertaken to interview you I have also given some attention to the practical aspect of the question. What do you think of having a go at astronomy? (*The Secret Agent*, 38)

regional authorities to which they had been relegated for centuries. Accordingly, this process of conversion is best explained by the concept of convention. Peter Gallison argues for a tripartite definition that encompasses the merging of mathematical science and international diplomacy in a watershed of consensus for the universality of weights and measures:

> The concept of convention widened, condensing into a word of triple resonance. *Convention* invoked the revolutionary Convention of Year II that introduced the decimal system of space and time; *convention* designated the international treaty, *the* diplomatic instrument that the French, more than any other country, pushed to the fore in the second half of the nineteenth century. More generally, *convention* is a quantity or relation fixed by broad agreement.[28]

For all the vaunted cooperation and scientific idealism of the international treaties, however, individual nations remained the principal bodies for disseminating and enforcing these relations. This would also be a critical issue in the rise of postal conventions that established sovereignty based on rights of distribution and maintenance of postal branches beyond national borders.

Even before its participation in the Prime Meridian Convention, Japan would for all external appearances begin keeping time with the West with its adoption in 1873 of the Gregorian calendar and twenty-four-hour clock. No doubt it was expedient for commercial, military, and diplomatic purposes to synchronize with the Western powers, which in an ideological sense the Convention simply reaffirmed. The shifts toward equivalence and commensurability with the timekeeping of the West were issued in government edict on December 9, 1872. In his recent work on the conceptions and conventions of time in Meiji Japan, Stefan Tanaka outlines the five rescriptings of calendrical and clock-time promulgated by the edict:

> 1. The abolition of the lunar calendar and the adoption of the solar calendar will occur on the third day of the twelfth month. That day will be January 1, Meiji 6 [1873].

28. Galison, *Einstein's Clocks, Poincaré's Maps*, 92.

2. The year will be divided into 365 days, with twelve months and an intercalary day every four years.

3. The keeping of time had been divided into day and night with each having roughly twelve hours. Hereafter, day and night will be equal, and a clock (*jishingi*) will determine the twenty-four units. The period from *ne* (rat) *no koku* to *uma* (horse) *no koku* will be divided into twelve hours and called *gozen* (morning); the period from *uma no koku* to *ne no koku* will be divided into twelve hours and called *gogo* (afternoon).

4. The telling of time [literally, ringing of bells] shall be in accordance with the schedule below. When asking about the time of a clock we have used *nanji* [the character for time (*ji*) is *aza* (section of a village)]; this will change to *nanji* [using the character *toki* (time)].

5. Days and months of all festivals will be adjusted to the new calendar.[29]

It is difficult to overstate the abruptness these changes sought to impose on the majority of the population. Notwithstanding the limited presence of Dutch-imported clocks and watches in the eighteenth century, as Timon Screech qualifies,[30] time was never conceived as an independent scientific variable divorced from nature, classical and religious texts, or the accumulated layers of local experience. Ringing of bells, burning of incense sticks, and the use of sundials and water clocks did not express an abstract representation of time, but a dynamic relationship to the human senses and the activities of the elements.

In the transition from the twelve two-hour blocks of the water clock of pre-Meiji epochs to the twenty-four-hour mechanical clock (whose own

29. Tanaka, *New Times in Modern Japan*, 8. The translation is Tanaka's.
30. Screech explains:

The symbolic force of the discovery of permanent, visible time cannot be overstressed. Even if watches were not particularly reliable (as many were still not), they nevertheless revealed an absolute temporal progression, and the vector seemed less of progress than towards an abyss. The connection of time-telling to the mind-set of the later part of the eighteenth century was already recognized by those looking back from early in the following century. "No one," wrote Ōta Nanpo in 1820 of the situation in the 1780s, "was without his imported European clock slipped in at the breast." Clock-time too was an imported system of representation. (*The Shogun's Painted Culture*, 98)

twelve-hour overlapping for day and night were enough to drive proponents of standard time to distraction), there was more at stake than a
change in basic measurements and their accompanying terminology.
Although *aza* 字 has an identical graphic representation to the Chinese
character for "letter" (字 *ji*), it is better thought of here as an indexical
marker or notation, in this case, referring to the animals of the Chinese
zodiac. Tanaka explains that it is drawn from the smallest unit of the
system of measurement in the village system, meaning a subvillage, consistent with the hierarchical, heterogeneous, and intensely localized
spatial (and linguistic) divisions of Tokugawa society. The new conceptualization of time as a mechanical order ruptured the existing social order,
as Tanaka argues:

> The new word for time, *toki*, does not have the same spatial connotation
> as *aza*, and by being grounded in the temporality of the solar calendar, it
> became affiliated with an abstract and mechanical system. Moreover, be-
> cause it was adopted at the same moment that a progressive developmental
> time was being implemented, this new reckoning of time was connected to
> a society oriented around what Koselleck calls a "horizon of expectations," a
> linearity where the future is some unknown better form rather than an
> ideal rooted in a previous world.[31]

Uprooted from the classical sources of authority and social relations of
the Edo period, temporality was no longer rectified by its continuities
with the idealized, imagined voices of the past, but synchronized to standardized, bureaucratic, and of course, national-imperial mechanisms.
The overlapping graphic representations of time are adequately conveyed
in a series of thin pamphlets issued by the Meiji government in the 1870s,
which allayed the confusions of a populace in transition by patiently
mapping the astrological markers of the water clock onto the Roman
numerals of a European clock face, along with a complete listing of the
revised dates for all major Shinto and Buddhist holidays, as well as days
of national observance (see figure 1.1).

The strictures of mechanical clock time provoked new shocks of the
modern as temporality was elongated or constricted in accordance with

31.   Tanaka, *New Times in Modern Japan*, 13–14.

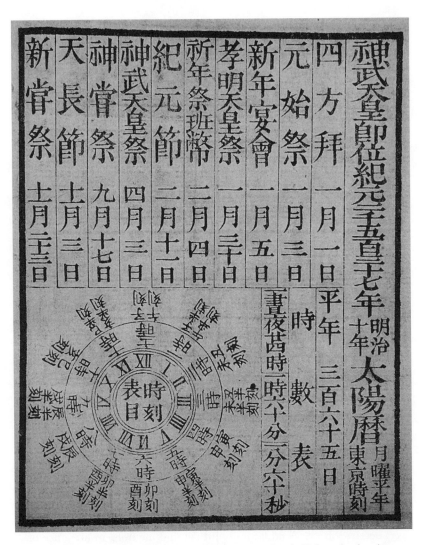

1.1 Calendar and clock time from *Meiji kunen taiyō ryakureki* (Solar calendar chart, 1876). Photograph by the author.

the time pressures of the business of empire, office work, and even the domestic bliss of the "high collar," if not yet white-collar, home. Kobayashi Kiyochika registers these reverberations in a political cartoon in the January 23, 1886, issue of the *Maru Maru Chinbun* 團團珍聞 (figure 1.2). In the upper half of this split-frame cartoon entitled "Enkōin no sumō"

袁航院の角力 (Sumo on the high seas), maritime friction in the Pacific between China and Japan is represented by sumo wrestlers squaring off for a match, a one-eyed ship named after major ports of call in each country substituting for each wrestler's head. Meanwhile, in the lower half of the split frame, "Jikan no hikinoshi" 時間の引きのし (Ironing time), a woman is ironing her husband's clock- and watch-patterned kimono. The English caption reads: " 'Time is money, but work is hard, I suppose, my love!' says the lady whose husband is an office worker."[32] The symbolic extension of work hours caused by smoothing out the wrinkles in the kimono cause the husband and another man, both dressed in Western business attire, to react with horror. Kiyochika conveys the metaphor of time as a fabric that can be contracted or stretched, made to conform to the abstract and mechanical dictates of the modern workplace.

Despite the radical changes of the 1872 edict, the system of official reign names (nengō) for marking periods of rule in pre-Meiji Japan was not so readily abandoned. A preexisting tradition that passed into the hands of the shogunate from the roughly two and half centuries of its political hegemony, reign names had been under the control of the shogunate and changed at irregular intervals to mark auspicious events or forestall inauspicious ones. Screech observes,

> In all parts of north-east Asia, time was built into blocks that were confined to the regimes that declared them, that is, they were geospecific. These blocks were known as "eras" (in Japanese nenkan). On the Continent, since the start of the Ming dynasty (1368), "eras" had been coterminous with the reign of the huangdi [emperor], but in the Tenka [shogun's realm] they were more culturally cadenced and promulgated as required in response to great events about every decade or so. Names were culled from classical texts to have a crisp, euphonic sound. Kansei, for example, which [w]as the era declared in 1789 (and lasting until 1800), came from an utterance of the sixth-century founder of the Chinese Sui kingdom: "when conducting government, do so with lenience."[33]

In other words, the Tokugawas' celestial mandate in the present was ratified through intertextual reference to classical texts, invoking their

32.   *Maru Maru Chinbun*, 16:510–11.
33.   Screech, *The Shogun's Painted Culture*, 99.

1.2 Kobayashi Kiyochika's political cartoon from *Maru Maru Chinbun*. Photograph courtesy of Waseda University Library.

precedent less as a guide than as an authorizing continuity—a mode of temporality entirely consistent with the poetics of the present, which harmonized multiple cyclical relations among seasons, famous places, and special occasions with suitable recombinant tropes. Starting with Meiji, the reign names reverted back to the imperial institution and were used to signify the duration of an "enlightened rule," the literal meaning of "Meiji," which ended only with the death of the emperor.

If it was unnecessary to explicitly confirm the revision to the *nenkan* system six years into the reign of the Meiji emperor, noticeably absent from the five articles in the December 9, 1872, edict is any mention of its nearest equivalent in the West, the Anno Domini system. The *nenkan* system continues to this day to impose a peculiar conditionality upon the logic of standardization: it is a series of arbitrary beginnings and uncertain ends. Given Japan's adoption of the twelve-hour clock and Gregorian calendar, it was inevitable that the Anno Domini system would come into contact and conflict with the chronology of imperial reign names. Akin to calls for the phonetic rescripting of national language from the 1870s that threatened to break the continuity with premodern writings and their textual connections to the imagined communities of classical antiquity, Anno Domini constituted another dangerous supplement to Japanese modernity that went to the heart of the emperor system's political and religious legitimacy.

As went the temporal, so too went the spatial. The metric system was originally calculated as a globally derived standard of weights and measures, with the meter that came out of the French Revolution in 1799 claimed by its advocates to represent with exactitude one ten-millionth of a quarter-arc of the Earth. Yet the salient feature of the version of the metric system put forth by the French at the international Convention of the Meter in 1875 was not the purported accuracy of a meter arrived at almost a century earlier, but use of the decimal and its divisibility into units of ten allowing for the widest range of commercial and scientific applications.[34] Although unevenly distributed in practice, the metric system was, improbably enough, endorsed by the United States. It was accepted first as one workable system among many as early as 1866, then with the importation of the metric standard from France in 1890 as a stopgap intended to help systematize and regulate the tremendous variety of weights and measure that prevailed from one locality to the next.[35] For its part, the Japanese government introduced the metric system in 1891, but only gradually phased out the *shakkanhō* traditional units of

34.   See Galison, *Einstein's Clocks, Poincaré's Maps*, 84–87.

35.   Michael E. Ruane's "100 Years of Setting Standards" brilliantly captures the sense of sacrality and secrecy which today attend the rare unveilings of the noncirculating, authorizing body at the center of the United States's crypto-metric system called the "National Kilogram":

> Human hands may not touch the object, lest it become altered in the slightest nano-way, and the air inside is filtered of all but dust particles a half-micron in diameter, or less. . . . The thick slice of platinum-iridium alloy, crafted by French instrument makers about 1889, [is] the only solid object in the country still used to define a measure. It's the National Kilogram. Primal, untainted by dirt, lint, the weight of a fingerprint, the calamity of a smudge. The immaculate kilo. The ultimate standard against which lesser kilos across the land may be compared. The official foundation, born of the French Revolution, on which the peculiar American system of ounces, pounds and tons is built. The cosmic sire of every bathroom scale in the country. (*Washington Post* April 1, 2001, F01)

measure.[36] While vestiges of the older systems remain today—floor plans calculated by the number of tatami mats is a case in point—it is important to remember that even these ostensibly indigenous measures have been standardized according to precision measurements and brought into equivalence with metric and English units.[37] It is at once standardization with the international conventions *and* the illusory continuity with the "national" past that paradoxically defines this aspect of Japanese modernity.

The rationalization of time and units of measure leads us to the strikingly similar reformist positions of Alexander Graham Bell and Tanakadate Aikitsu (1856–1952), who actively sought to bridge the disparate relations of language, scientific measure, and modern writing machines. Alexander Graham Bell went before the Committee on Coinage, Weights and Measure in the House of Representatives in 1906 to advocate the adoption of the metric system in the United States. The tenor of his remarks is set forth in the title of the transcripts originally published in the March 1906 *National Geographic*: "The Metric System: An Explanation of the Reasons Why the United States Should Abandon Its Heterogeneous Systems of Weights and Measures."[38] Not only is there an endemic lack of accuracy in calculating agreed upon units of measure throughout the nation, Bell maintains, but sometimes two or more definitions of units are used inconsistently in a given industry or geographic region. The chairman of the committee concurred with Bell's assessment, noting that even the U.S. Mint employed no fewer than four official systems of measure: avoirdupois, troy weight, apothecary's, and metric. Another advantage of the metric system Bell praised is that the decimal obviates the need for complex calculations, its homogeneous units doing away with unwieldy conversions across pounds, ounces, gallons, quarts, and so on. Bell insisted there was an additional linguistic advantage. Since

---

36. The metric system underwent several stages of legal ratification in 1921, 1951, and 1956. In principle, old measures were officially banned in 1966.

37. Standard equivalents for Tokugawa era weights and measures today run as follows:

1 *sun*	= 3 cm. or 1.2 inches	1 *ken*	= 1.7 m. or 5.5 feet
1 *bu*	= 3 mm. or 0.12 inches	1 *ri*	= 3.9 km. or 2.4 miles
1 *shaku*	= 30 cm. or 1 foot	1 shaku	= 180.4 liters or 47.7 gallons

38. Reprinted in Bell, "The Metric System."

its etymology is self-explanatory in the majority of European languages, the universality of the metric standard means that all nations will speak, as it were, a common language of measurement. This, too, would obviate the need for translating across disparate linguistic units, saving time and money for all parties involved. Although Bell was ultimately unsuccessful in persuading the federal government to abandon its antiquated system, he might take a small measure of comfort in knowing that the metric standard would gradually become the indispensable, if spectral, supplement to America's national system of measure.

Akin to Bell in his commitment to scientific research and promotion of phonetic reform, Tanakadate was a polymath who made substantial contributions to fields from electromagnetism and seismology to aviation.[39] As a lifelong advocate of standardized measurement and for his scientific achievements, Tanakadate was selected for the standing committee sent by Japan to the 1907 Paris Conference on the Metric System of Weights and Measures, which resulted in the passage of a reform bill by the National Diet later that year.[40] Moreover, he applied his knowledge of metrics to language early in his career when he invented the phonemically based Nippon-shiki, or "Japanese system" of romanization in 1881, which, unlike James Hepburn's Hébon-shiki from 1859, prevented loss or overspecification when mapping to and from kana. Kida Junichirō remarks that the inspiration for Tanakadate's efforts to convert the Japanese script came from his experience in college witnessing the audio-visual spectacle of romanized Japanese text read aloud, inscribed, and relayed back via an early model of the phonograph: "Tanakadate Aikitsu, known for his accomplishments in geophysics and the systematizing of weights and measures, had marveled during his student days at Tokyo University when a British instructor read a Japanese text written in *rōmaji* and recorded it on a wax cylinder phonograph. Tanakadate subsequently began writing most of his scholarly treatises and correspondence in *rōmaji*."[41] An outspoken critic of the ideology of "Eastern spirit, Western

39.   Tanakadate studied electricity and magnetism in Glasgow with William Thomson, the Lord Kelvin, from 1888 to 1890. In addition to his contribution to the science of measurement, Thomson invented the mirror galvonometer telegraph and reengineered the long distance wires used in the first successful transatlantic cable in 1866.

40.   See, for instance, Tanakadate's *Metoru-hō no rekishi to genzai no mondai*.

41.   Kida, *Nihongo Daihakubutsukan*, 109. Translation is my own.

technology" (*wakon yōsai*) that prevailed within the nationalist rhetoric of Civilization and Enlightenment, Tanakadate not only promoted romanization during the late 1880s and early 1890s when the Romanization Society and Kana Society were at the peak of their popularity, but continued to advance its cause after the unified style came to dominate the debates over language and script reform. As Kida recounts, Tanakadate was a financial backer for the publishing company Nippon-no-Rōmaji-Sya (Japanese Romanization Society) founded in 1909, and used his position in the House of Peers to which he was appointed as a representative of the Imperial Academy in 1925 to deliver an annual speech written in romanization.

Tanakadate also composed modern *waka* (5-7-5-7-7 verses) that praised the almost boundless possibilities for romanization, provided the disparate languages and writing systems of the world could be made to pass through the bottleneck of the alphabetic signifier:

Many are the nations/Whose grass-blades of words grow thick/Parted so easily/by the letters of *rōmaji*

Kuni wa ōku koto no hagusa wa sigekumo humi wake kayō moji wa rōmaji[42]

Similarly we see in the poem below tropes reminiscent of the phonocentric aspirations of Native Learning scholar Motoori Norinaga, in which Tanakadate deftly transports the transcendental spirit of language into universalizing scripts and technologies, bringing Japanese into a linguistic order akin to the global uniformity of the metric system and standard time:

The prosperous path to the spirit of words/steadily comes into view/The pleasure of striking [the keys of] the typewriter

Kotodama no iya sakae yuku miti miete taipuraita no tataku tanosisa[43]

Tanakadate dissolves the conventional tropes and compositional processes for Japanese poetry (such as flowery language composed on the tip of a

---

42. Ibid., 112. I have provided the romanization here according to the Nippon-shiki system. Translation is my own.

43. Ibid., 112. Translation is my own.

brush; *tanka* as a poetic genre of national particularity), replacing them with an internationalist, machine-driven milieu.

Whether in terms of spatial, temporal, or linguistic changes, standardization is a critical basis for modernity's emergence. It follows that before we can begin to discuss technology in terms of writing machines and apparatuses, we must be attuned to the organizational principles and techniques that preceded them, including as their very grounds the Heideggerian concept of standing-reserve. Our contemporary outlook on the instrumentality of technology is indisputably a product of changes that took place since the late nineteenth century, yet is paradoxically beset by a frequent tendency toward the erasure of media history. I have attempted to sketch out in this chapter how the wide-ranging discursive transformations that produced modern subjectivity in the West were also present in Meiji Japan. The ensuing two chapters in part I track the medial conditions responsible for sending and receiving messages—namely, those contained within the postal network. The rise of the postal system, including telegraph and telephone, provided the wiring of the modern nation-state and its stark contrast with the ancien régime of the Tokugawas.

CHAPTER 2

Telegraph and Post

Maejima Hisoka

Although some form of courier service had existed in the West for centuries,[1] it was only in the 1840s that England, France, Germany, and the United States established modern postal systems to collect, sort, and deliver all forms of mail. In the United States, the use of postal stamps began in 1847, followed by the use of mail-drop boxes on city streets in 1854, and universal free delivery arrived in most major towns and cities by the 1860s. The Japanese post was therefore not that far behind schedule when Maejima Hisoka (1835–1919) was appointed Japan's first postmaster-general in 1871.[2] Maejima was chief architect of a postal revolution that, within the span of three or four decades, replaced the shogunate's "fleet-footed" (*hikyaku*) courier service[3] with a national-imperial system on par with any in the world. In addition to the delivery of the

1. For a useful overview of premodern Western postal services and discourses, see Holzmann and Pehrson's *The Early History of Data Networks*, 1–44.
2. For a brief overview of Maejima's role in the establishment of the Japanese postal system, see MacLachlan's *The People's Post Office*, 35–52.
3. As chapter 3 explores in greater detail, the Tokugawa system of relay-runners carried important letters and documents along five main arteries leading to Edo, the Tōkaidō, Nakasendō, Kōshū Kaidō, Nikkō Kaidō, and Oshū Kaidō, with horses and riders in the service of the shogun (*denma*) also provided when necessary. Privately run services called "town couriers" (*machi-hikyaku*) were also used by merchants for their own purposes along these official routes.

mail and administration of the telegraph and telephone, which were eventually assigned to a separate government ministry,[4] Maejima introduced savings accounts, postal orders, and other instruments of monetary exchange. He is popularly credited with coining the Japanese terminology for the postal system (*yūbin*), the postage stamp (*kitte*), and so forth. Notably, Maejima also participated in the some of the earliest efforts to standardize weights and measures, two years prior to the Meiji government's edict on time in 1873.

Maejima was to successively occupy prominent positions of leadership in Meiji industry, government, and education. After serving at the helm of the postal service, Maejima became the second president of the Tokyo Senmon Gakkō, the forerunner to Waseda University (1886–1890), was made a baron and appointed to the House of Peers (1904–1910), and took a leading role in financing rail development in the Hokuriku region in northeastern Honshū. As postmaster-general he also presided over the creation of the *Yūbin Hōchi* 郵便報知 (Postal report), a progressive newspaper that helped cultivate the talents of leading language and script reformers, including Yano Fumio (pen name Yano Ryūkei) as its editor in chief, and the talented young shorthand reporters Wakabayashi Kanzō and Sakai Shōzō.

I start with this biographical focus on Maejima less to recap the contributions of one of the "great men of Meiji" than to expand beyond the narrow historical role he is typically made to serve in Japanese literary studies. He is best known for his *Kanji gohaishi no gi* 漢字御廃止之儀 (Proposal to abolish Chinese characters) submitted to the shogunal authorities in 1866 while still an instructor at the Kaiseijo, the shogunate's school of Dutch Learning. It was a brash missive that went unanswered even as the shogunate was faced with an unprecedented crisis of foreign

---

4. Official terminology for the postal system changed several times during the initial decades of transition, reflecting the redistribution of authority and labor as the Meiji state grew in size and complexity. As Maejima himself notes in a late Meiji English publication, "Up to 1885 the entire postal system was under the control of the Transportation Bureau (*Ekitei Kyoku*); but in that year the Communications Department (*Teishin Shō*) was created, and the Post and Telegraph Services, having been united were put under its jurisdiction. . . . In 1900, with the consent of the Imperial Diet, a postal service was enacted, and the system became a complete organization" (Ōkuma, ed., *Fifty Years of New Japan*, 2:409).

incursions and teetered on the verge of collapse. In spite of these exigencies, it is today widely regarded as the opening salvo in the reform of Japanese script away from the heterogeneity of Japanese écriture and toward the hegemonic use of the unified style. For instance, it is the first of 169 excerpted documents that make up the six stages of development in language historian Yamamoto Masahide's *Kindai buntai keisei shiryōshū* 近代文体形成資料集 (Source materials for the formation of a modern writing style). Although he was by no means the only scholar engaged in this endeavor, Yamamoto was principally responsible in the late twentieth century for defining and elevating the history of a triumphant unified style.[5] Regardless of these popularly held views of its uniqueness, we must refute the idea that this sort of phonetic turn was unique to Japan. It is also necessary to more rigorously scrutinize where Maejima has been given more credit than he is due as a language reformer, for instance as the harbinger of phonetic transparency portrayed by Karatani Kojin in *The Origins of Modern Japanese Literature*, serialized during the same period as Yamamoto's canonization of the unified style. Moving beyond the contributions of Maejima's and other human hands, the latter two sections of this chapter investigate the networks of post and telegraph. In the process of establishing new forms of control and communication inside Japan proper and its colonies, the postal system also achieved a high degree of visibility across disparate media and genres. By recuperating its media history, we also come into a better understanding of this episteme.

Maejima's proposal was not made public until 1884, when the newly established Kana Society reissued it. Fittingly, Konishi Nobuhachi (1854–1938), an associate of Isawa Shūji's and a prominent figure in his own right for language reform and education for the deaf and blind, wrote a kana-only preface that praised Maejima's efforts.[6] Konishi credited him for founding the modern postal system (*yūbin seido*) and introducing a new

---

5. Many factors were at work, including the postwar loss of empire and teleological retrenchment of Japanese national particularity in ethnolinguistic terms; the tight correlation between modern literary style and the heights of linguistic expression; and the renewal of standardization efforts in postwar compulsory education and mass media such as radio and television.

6. Konishi served as principal of the Tokyo School of the Deaf (*Tokyo mōa gakkō*) from 1893 to 1910. The school taught the Japanese Braille system adapted by one of the school's instructors, Ishikawa Kuraji, in 1890.

instrumentality to Japanese "letters," no doubt well aware of the dual resonance the word holds in English.

The corpus of mid-eighteenth- to early nineteenth-century writings on alphabetic writing by earlier Dutch Learning scholars notwithstanding,[7] Maejima's treatise was the first attempt to explicitly encourage a national education policy along phonic guidelines. Maejima forcefully argues that the proper function of the state is the "normal education" (*futsū kyōiku*) of the people—a position consistent with efforts in Europe and America to conjoin state-sponsored national education with language standardization. Almost two decades before Fukuzawa Yukichi's "De-Asianization" 脱亞論 (1885), Maejima tied the reform of writing and the promulgation of national education to the need for Japan to achieve parity with Western powers and divorce itself from the spiritual decadence of a contemporary China humiliated by the Opium Wars and trade concessions. Yet Maejima also called for severing associations to the classical Chinese learning promulgated by the shogunate's neo-Confucian ideologues. He decried the waste of time spent studying Chinese texts and insisted on practical Western studies written in kana, which would have the positive benefit of revitalizing "Japanese spirit." These layers of anti-Chinese sentiment did not mean Maejima claimed kana represented an essential Japaneseness on the order of a latter-day disciple of Native Learning, however. His justification for the superiority of kana as the national standard lay in its status as a phonetic script akin to the alphabet. His proposal was thus intended to break the monopoly of Chinese as a writing system and

---

7. Twine identifies several precedents regarding the relative advantages of alphabet writing over Chinese characters going back to the mid-eighteenth century:

Gotō Rishun (1702–1771), in *Orandabanashi* (Tales of Holland, 1765), wrote out and briefly described the Dutch alphabet; Ōtsuki Genpaku (1757–1827), in *Rangaku Kaitai* (A Guide to Dutch Studies, 1783), remarked on how easily it could be learned; Shiba Kōkan (1747–1818), in *Oranda Tensetsu* (Tales of Holland, 1796), praised the ease afforded reading by the use of a phonetic script. Shiba suggested that Chinese characters be replaced by kana, an idea supported by Yamagata Bantō (1748–1821) in *Yume no Shiro* (The Value of Dreams, 1802), and Honda Toshiaki (1744–1821) in *Seiiki Monogatari* (Tales of the West, 1798). Honda even recommended the use of the Western script itself, which, he noted, was more flexible than kana and had the advantage of being internationally recognized (*Language and the Modern State*, 225).

dislodge the ideogram (*keishō moji*) as a determinant of cultural prestige and belonging in the modern world.

Karatani has argued that Maejima's promotion of kana was nothing less than a desire to rescript Japanese national identity in correspondence with the phonetic transparency of the West. An important caveat must be offered, however. Although Karatani's words ring true for later Meiji language reformers such as Mori Arinori, Isawa Shūji, and the various proponents of shorthand, the case for phonetic transparency was never made by Maejima himself. Given his superlative education and coming of age prior to the standardization of national language, Maejima was of course keenly aware that kana were nothing more or less than simplified versions of Chinese characters. His proposal to abolish Chinese characters does not begin to address the standardization of kana, which in Tokugawa popular culture often had a bewildering array of possibilities. The closest he comes to a new logic of kana was to recommend punctuation and spacing, which are essentially typographic distinctions, to avoid a confusion that did not exist with Chinese characters. By the same token, whereas Karatani is correct in broadly asserting that Maejima's "conception of spoken language was itself rooted in a preoccupation with phonetic writing,"[8] it is another matter entirely to extrapolate that "once this view had been established, the question of whether or not *kanji* were actually abolished became moot. Once even Chinese characters had come to be seen as subordinate to speech, the issue became simply a choice between characters and the native phonetic syllabary."[9] In actuality, before we can begin to think of the unified style as consisting of Chinese characters and kana, which were always already a standing-reserve for Japanese écriture, it is necessary to consider the many experiments with phonetic scripts and mechanical codes from the 1870s to around 1900, when Chinese characters, kana, and, at least in principle, Roman letters, were becoming standardized. I will take up the further implications of the debates over "the question of national language and scripts" in part II. For now, I wish to return our focus to the discourse networks of telegraph and post that Maejima had a hand in setting up, but that quickly took on a historical agency of their own as modern media.

8. Karatani, *The Origins of Modern Japanese Literature*, 47.
9. Ibid., 47.

Instant Messaging

One of the earliest writing technologies of the nineteenth century that embraced a simplified script written by mechanical arms is also perhaps the least known. The optical or semaphore telegraph, such as the prototype invented by Claude Chappe in 1791–1792, rapidly communicated across distances through encoded sequences of an armature built onto towers or high promontories (figure 2.1).[10] As Tom Standage explains of the revolutionary coding of Chappe's semaphore telegraph,

> The design allowed for a total of ninety-eight different combinations, six of which were reserved for "special use," leaving ninety-two codes to represent numbers, letters and common syllables. A special codebook with ninety-two numbered pages, each of which listed ninety-two numbered meanings, meant that an additional ninety-two times ninety-two, or 8,464 words and phrases could be represented by transmitting two codes in succession. The first indicated the page number in the codebook, and the second indicated the intended word or phrase on that page.[11]

Chappe's telegraph, which he initially called a "tachygraph" or "fast writer,"[12] was set up to send encoded messages over long distances using a mechanical armature placed on promontories or a series of towers and read using telescopes—letters writ large in the sky at the dawn of the machine age. Although Chappe's invention would be superseded in the 1840s by the electric telegraph, the semaphore found renewed significance in the signaling systems of the railroad industry.

The electric telegraph that superseded the semaphore was defined by its capacity to collapse distance and time in the encoded relay of electric

---

10. For further reading on the diversity of telegraphic experiments, including Chappe's "Synchronous System" using pendulum-driven clocks and the alternative panel or shutter telegraph, see Holzmann and Pehrson's *The Early History of Data Networks*, 48–96.

11. Standage, *The Victorian Internet*, 10.

12. A name that was later coopted by one of the competing systems of shorthand to Pitman's phonography, and the literal translation of the Japanese term for shorthand coined by Yano Ryūkei in 1883, *sokki* 速記.

2.1 Chappe's Semaphore Telegraph, c. 1792. Photographed in Musée des Arts et Métiers.

impulses across wired networks. Preceding the often fraught negotiations between ideologies of standard national language and phonetic script reform, the telegraph precipitated a no-less thorough reconceptualization of the metaphysics of presence by mechanically outstripping the monopoly of instantaneous transmission presumed to reside in the spiritual outflow from thought to breath to voice.

Indeed, the revolution in linguistic science under way in the 1890s took place amidst the reenvisioning of language in terms of the precision and fidelity of signal processing. Language need no longer be restricted to the positivistic relation of words to things or to the phonocentric premise of speech over writing, but could be reconceived as codes sent and received by telegraph. Timothy Mitchell offers the following remarks by Saussure's colleague at the Collège de France in regard to Marconi's electromagnetic experiments, patented in 1896, that eliminated the wire previously needed to link transmitter and receiver:

> "Words are signs," it was now declared. "They have no other existence than the signals of the wireless telegraph." This claim was made in 1897 by Michel Bréal, professor of comparative grammar at the Collège de France. The significance of arguing that words were mere signs, as empty in themselves as telegraph signals, was that a language could now be thought of as something more, existing apart from words themselves. The meaning of a language existed not in the plenitude of words, which were arbitrary marks meaningless in themselves, but outside them, as a semantic "structure."[13]

It is worth pointing out, however, when Saussure set out to establish the conditions of interpersonal communication irrespective of media or other contexts he nevertheless labeled it "the speech circuit."[14]

Well in advance of these recognitions of the material conditions of language by Europe's most prominent linguists, the gist of this transformation in the modes of sending and reconceptualizing writing had already

13. Mitchell, *Colonising Egypt*, 140. Marconi's wireless telegraph, the forerunner to the radio, was sponsored by the British Post in the 1890s.

14. Saussure asserts, "In order to identify what role linguistic structure plays within the totality of language, we must consider the individual act of speech and trace what takes place in the speech circuit. This act requires at least two individuals: without this minimum the circuit would not be complete" (*Course in General Linguistics*, 11).

been made abundantly clear by Commodore Perry's expedition to open Japan to trade in 1853. Accompanied by the quarter-scale railroad train, the Americans used a telegraph that demonstrated instantaneous speed coupled with multilingual interchangeability:

> A piece of level ground was assigned for laying down the circular track of the little locomotive, and posts were brought and erected for the extension of the telegraph wires. . . . The telegraphic apparatus, under the direction of Mssrs. Draper and Williams, was soon in working order, the wires extending nearly a mile, in a direct line, one end being at the treaty house, and another at a building expressly allotted for the purpose. When communication was opened up between the operators at either extremity, the Japanese watched with intense curiosity the *modus operandi*, and were greatly amazed to find that in an instant of time messages were conveyed in the English, Dutch and Japanese languages from building to building.[15]

At the outset of its arrival in Japan, then, the telegraph held no official relation to any national or international language, but was theoretically open to any system that could be suitably encoded. Romanization, kana, and even kanji were equally plausible in this showcase of the telegraph. Although the records of the expedition do not specify what form of "Japanese" language was employed, it would be misleading to assume the linguistic transparency and one-to-one exchange value between letters and sounds that obtain today, which are a legacy of national language standardization that followed the advent of the telegraph and other writing machines.

Unlike the telegraph lines that typically went up alongside the railroad in Europe and North America, the development of a Japanese telegraphic industry in early Meiji quickly outpaced the development of train lines. In 1869, three years before the inaugural Shinbashi-Yokohama train line, Tokyo and Yokohama were connected by telegraph, and an undersea wire connecting Nagasaki and Shanghai was put into service in 1871. The main southwestern line from Tokyo to Nagasaki was completed by early 1873, and the main northeast line between Tokyo and Aomori was

---

15. Hawks, *Narrative of the Expedition of an American Squadron to the China Seas and Japan*, 357.

finished in 1874. The exponential growth of the telegraph provides some quantitative measure of the revolution in technologies of writing that prepared the way for further national development. Looking back at the humble beginnings of telegraphy in 1909, Ōkuma Shigenobu would observe: "In 1871 the number of domestic telegrams totaled only 20,000, but in 1907 messages sent and received rose to 24,413,965, not including Formosa. The figures stood at 26,000,000 when those for foreign messages were added."[16]

Despite the slow growth of Japan's rail industry, the colonial exclusion here is telling. At the same time that wires were abuzz throughout the domestic interior, they were also stretching to cover the separate and unequal territory of Japan's East Asian colonies. Yet the aggressive assimilationist policies that began on the archipelago with the consolidation of rule over Hokkaidō and Okinawa in 1869 were not yet matched by efforts to accommodate the ethnically, linguistically, and geographically more distant colonies into the imagined community of Japan proper. Elementary school language and geography books published by the Japanese Ministry of Education after 1895 reveal that Japanese children were taught to recite the now *five* islands of the Great Japanese Empire in descending order from north to south: Hokkaidō, Honshū, Shikoku, Kyūshū, and Taiwan. With the exception of the imperial universities and schools for elites, linguistic and educational campaigns for the Taiwanese did not begin in earnest until the 1920s. By contrast, Hokkaidō was rapidly incorporated into the national imaginary as Japan's wild frontier even as its cities grew and its lands were wrested from the aboriginal people.

We might consider the ease with which Doppo could refer to Hokkaidō as native soil and Japan's frontier in the late 1890s alongside the writings of one-time telegraphic engineer Kōda Rohan. Kōda spent the period from 1885 to 1887 stationed in the village of Yoichi, near present-day Otaru, in return for his education at the state-run Telegraphers Training School, and witnessed firsthand the colonization of the northern territory. Kōda's term of duty coincided with the laying of telegraphs lines and railroad tracks across the north. The pamphlet "Outlines of the History of Telegraphs in Japan," published in English by the Japanese

16. Ōkuma, ed., *Fifty Years in New Japan*, 2:420.

Ministry of Communications (Teishinshō) for the 1893 World Exposition,[17] records the rapid progress that helped to colonize the island previously known in Japanese as Ezo, now renamed Hokkaidō, and set Tokyo as the central communications hub on Japan's still *four* main islands:

> [In 1874–75], the Kaitakushi (the local government of the island of Yezo [*sic*]), after making application to the central government, established the Hokkaidō lines; of these one line runs from Hakodate, in the province of Toshima, to Sapporo in the province of Ishikari, passing through Otaru in Shiribeshi, and the other line runs from Sapporo to Muroran in Iburi.[18]

Kōda was also only too well aware of the myth of the empty lands applied to Hokkaidō, much as the Americans had done with their western frontier. As Robert Sayers affirms, "Hokkaido was incorporated in the new Meiji state and its 'vacant' lands opened up for settlement. Facing an influx of farmers and other settlers, the Ainu retreated ever further into the interior."[19] Although Kōda abandoned his post to return to the literary fold in Tokyo, in 1889, he retrospectively wrote the short story "Yuki funpun" 雪ふんぷん (Snowflakes dancing), which idealized the struggle of the Ainu against the colonialist Japanese.

In the same year that Kōda was dispatched to Hokkaidō, the telegraph was cast in a satirical light in the *Maru Maru Chinbun*, where it was used to jab at the unequal treaties with Western powers inherited from the shogunate and not fully repealed until 1899. Inveighing against the conditions that have made the treaty ports across Japan and even Tokyo into semicolonized "domestic mixed settlements" (*naichi zakkyo*), the bilingual editor's column[20] from September 19, 1885, describes a new

17. The copy in the Cornell University Library bears the inscription: "The Gift of the Imperial Japanese Commission, World's Columbian Exposition, 16/2/93."

18. Mayeda, "Outlines of the History of Telegraphs in Japan," 27–28.

19. Sayers, "Ainu: Spirit of a Northern People," 879.

20. The journal includes side-by-side *"wakan"* columns in kanbun, colloquial rakugo scripts, and articles written in mixed kanji-kana scripts. Like the bilingual editor's column begun in 1885, political cartoons were also captioned in Japanese and English. On the transitions to genbun itchi in its editorials, see Yamamoto's *Genbun itchi no rekishi ronkō*, 238–244.

variety of telegraph, a table-turning device to spread the word about the arrogance and chauvinism of the American-led coalition of occupiers:

> Some Yankee rascal once said to us, that he had invented a new telegraph; and to our astonishment his proposition was to place a line of women fifty feet apart and commit the news to the first one as a very profound secret. Now, our Great City of Tokio [*sic*] will soon be provided with those living telegraph-poles, as the treaty Revision that has been going on for so long a time will have to be finished anyhow, when the fair angelic maidens, such as we have seen in a dream, as well as the filth and scum of the Western civilization will come in to Americanize or Europeanize our Great City and holy country more completely.[21]

The telegraph thus provided an effective means of critiquing of Western imperialism, even as it was being used to make inroads for Japan's colonial expansion into Okinawa and Hokkaidō.

Expressions of righteous indignation over colonial, racial, or gender inequalities notwithstanding, the telegraph would remain a multilingual enterprise. All things being equal as costs were set per message according to distance, prices for messages composed in Roman letters were set by the number of words (*go*), while those for Japanese messages were set by letter of katakana (*ji*). Moreover, postal law from the 1870s until at least the mid-1880s would briskly state, "The Telegraph lines are worked conjointly by Europeans and Japanese. The latter act, under Foreign superintendence, as Operators, Clerks, Engineers, Inspectors, and Linemen."[22] Essential equipment such as the Morse Printer, the Bréguet Alphabetical, and the single-needle instrument were likewise imported from the United States and Europe until the mid-1880s, when Japan began its own industrial production of telegraphic equipment. Students in the newly formed telegraphers' schools, such as the indigent Kōda, were educated not only in the use of the mechanical apparatus, but also in English and French. When sufficient numbers of Japanese had been trained in all levels of operation, the foreigners would eventually be phased out (as was

21. *Maru Maru Chinbun* 16 (1885): 156.
22. Yūseishō, eds., *Yūsei hyakunenshi shiryō*, 19:36.

similarly the case of the foreign hired hands in universities and schools), yet the demand for interchangeable communication in major European languages continued unabated. The invective of the *Maru Maru Chinbun* editors notwithstanding, technologies of writing in the service of global commerce and imperialism would last far longer, and with more pervasive cultural effects, than the onus of actual mixed settlement during the unequal treaties period.

If the telegraph was the consummate media technology for generating the shortest paper trails over the longest distances, the modern postal service remained for the better part of the next century the comprehensive, institutional monopoly for sending and receiving writing. Since postal systems coordinate material flows of information, but are not necessarily revealed in them, we must look once again to external auditing to engender their visibility. Regrettably, many prints and photographs from the period focus exclusively on the modern architecture of office buildings and the hoopla of the crowds. A classic example is Hiroshige III's often-reprinted *Yokohama yūbin-kyoku kaigyō no zu* 横浜郵便局開業之図 (Illustration of the inauguration of the Yokohama Postal Office) from 1874, which celebrates the exterior of the buildings that house the post and telegraph. We should turn our attention instead to Kubota Beisen's watercolor series *Yūbin gengyō emaki* 郵便現業絵巻 (Illustrated scrolls of postal work), which was commissioned for the 1893 Columbian Exposition in Chicago and put on display the multiple stages and labor divisions of gathering, sorting, distributing, and delivering the mail. They remind us of the innumerable small details that marked transitions to the modern postal system. For instance, uniforms are not yet completely uniform: there is a mix of Japanese- and Western-style costume, men in trousers and coats jostling with men, women, and children in kimono and hakama in the lobby of the Tokyo Post and Telegraph. A mixture of transportation methods from rickshaws and runners to horse-driven carriages, steamboats, and trains would likewise appear here and elsewhere in prints of the era displaying the postal service. As one ranges over these images, questions inevitably arise about the function of the state in regulating, delivering, and, inevitably, reading, the mail. The postal service as an outpost of empire also reemerges through the implementation of national and international conventions.

Secret Correspondences

It is easy to overlook the seemingly ordinary processes of modern life in the face of the quaint and exotic practices of yesteryear. Yet the second decade of Meiji was already a far cry from the days of fleet-footed runners and two hundred or so wooden letter collection boxes distributed along the post roads of the Tokugawa era. The vast proliferation of mailboxes now served as unmanned outposts in a system of standard postage and increasingly interconnected global postal networks. Before these processes fade into the background of modern social formation, it is possible to see in them the near-complete absorption of postal networks into the regime of state power. Although we take for granted that our communiqués will be read only by those for whom they are intended—flimsily safeguarded by the sealed envelope, folded paper, and designated address—the national postal service in essence amounts to the processing of private letters through public channels. The picture postcard offers one of the best examples of the contradictions solicited by the postal process. Despite the romantic overtones displayed by a woman anonymously mailing a letter to her lover in Sumiike Kuroteibō's picture postcard entitled "Secret Correspondence" (figure 2.2),[23] privacy was never anything more than a legal fiction, an enticement to allow oneself to be interpolated into the postal system. Even the medium of the picture postcard—privatized by the Meiji government in 1909—encouraged the public to send its mail sans protection of any sort. The irony of Sumiike's postcard is that it depicts a woman slipping a letter into a ubiquitous pillar postbox (with correct postage, one assumes) as if to afford the viewer the illusion of security from the prying eyes of neighbors and petty bureaucrats in the postal office. Still, Bernhard Siegert reminds us what lies behind this smooth transfer of messages is the potential for surveillance, interception, and seizure:

23.   *Himitsu tsūshin*, published in the *Kokkei Shimbun* 滑稽新聞 in 1908. The *Kokkei Shimbun* increased its sales with regular inserts of picture postcards in the latter half of the 1900s. Unfortunately, almost nothing is known of Sumiike, the pseudonym for a popular picture postcard artist for the newspaper.

2.2 Sumiike Kuroteibō, picture postcard *Himitsu tsūshin* (Secret correspondence) from the *Kokkei Shimbun*, 1902. Photograph courtesy of the Postal Museum Japan.

The ubiquity and invisibility of the state were thus to be found in the representation of the postal system as a medium for private correspondence between cognitive subjects. . . . As postal systems became a technology of the government with the invention of postage and the monopolization of service, the people likewise came to believe they were capable of determining their own affairs postally. Institutionally, this meant that the postal system fell under police jurisdiction.[24]

Accordingly, the police detective finds his unlikely counterpart in the figure of the postal inspector. Fodder for popular fiction since the nineteenth century, the intricacies of postal law and their evasion are no less a function of technological limits and the social contract that makes the post an extension of the state akin to any other public utility—it exists on behalf of the people so long as they do not interfere in its operations.[25]

Hegemonic surveillance works best when it is unobtrusive and has the full cooperation (or ignorance) of the community. This is readily borne out by the Meiji postal code, which in no uncertain terms lays down the law as follows:

Article 1. Mail will be managed by the government.

Article 2. No private person whatsoever shall make it his business to deliver mail.

. . .

Article 10. In handling mail, an act which is committed against the Post Office by a person without legal capacity will be considered as having been committed by a person of full legal capacity.[26]

24. Siegert, *Relays*, 9.

25. By way of analogy, we might consider the expansion of the telephone service roughly a decade later: when anonymity could more or less be afforded by direct dialing without the prying ears of network operators, telephone become another site for personal and sometimes illicit conduct; as Siegert reminds us in *Relays*, the late twentieth-century AT&T advertisements that invited callers "to reach out and touch someone" are a case in point. Nonetheless, whether it is a landline or mobile phone, the machines that facilitate contact are always under surveillance by telephone companies that keep logs of every call made—logs that can be turned over to the state as well as divorce attorneys.

26. Military Government Translation Center, trans., *Japanese Postal Laws*, 1–3. Precise date of publication is unclear.

Additional articles in the Japanese postal code enforce the postal imperative to deliver the mail on time, invoking as needed rights of passage and varying degrees of eminent domain. Article 4 states:

> When there is any obstacle on the road and it is difficult to pass, the mail dispatch carrier, mail carrier, horses or vehicles used for mail carrying purposes, while on duty, may pass through buildings, lands, rice fields, gardens, or other places which have no wall or fence around them. On such an occasion, the Post Office must make reparation for the damage at the request of the person suffering the damage.[27]

Evident from these postal laws, then, are the pathways of circulation and delivery as an exercise of state power and national identification—writing that knows no boundaries within the confines of the nation it serves, yet that may also be intercepted and confiscated by that same authorizing body.

Although it is hardly as well known as the verse that adorns post offices throughout the United States, "Neither snow nor rain nor heat nor gloom of night stays these couriers from the swift completion of their appointed rounds,"[28] Tsubouchi Shōyō's children's nursery rhyme, "Yūbinbako no uta" 郵便箱の歌 (The post box song) praised Japan's new postal empire. First published in volume five of the *Kokugo tokuhon jinjō shogakkō yō* 国語読本尋常小学校用 (Language primer for use in normal elementary schools), in 1900, its lyrics gently yet repetitively reinforce the *heimlichkeit* character of the post box in every neighborhood in the nation. The postbox is the hub of neighborhood goings-on, personified as a tireless messenger whose drop slot is constantly active, the postcards and parcels it receives and disburses connecting the nation with the thoughtful kindness of a favorite aunt.

---

27. Ibid., 2.

28. As Holzmann and Pehrson note, the lines are, in fact, an adaptation from Herodotus's descriptions of the ancient Persian courier system in *The History*: "And him [the messenger] neither snow nor rain nor heat nor night holds back for the accomplishment of the course that has been assigned to him, and as quickly as he may" (*The Early History of Data Networks*, 4).

The Post Box Song	Yūbinbako no uta
On every corner of the streets in the capital	Toshi no machi no yottsu kado no
The post boxes say:	Yūbinbako to iu yo wa
"Surely there are no others as busy as us	Satemo, isogashi, warera hodo
Nor as hurried.	Sewashikimono wa, mata araji.
From the crack of dawn 'til late at night	Asa wa hikiake, yofuke made
In they go and out they come, opening and closing,	Ireru, toridasu, sono aketateni,
Thump, thump, tha-thump	Potari, pottari, pottariko,
There is no rest for the mail drop slot.	Sashireguchi no yasumi nashi.
First there are the kinds of mail—	Mazu, yūbin no shinajina wa,
Sealed letters, print matter, and unsealed envelopes	Fūsho, Fūto, hirakifū,
Return-postcards, ordinary postcards	O-ofuku-hagaki, name-hagaki
The different scripts and people who send them	Moji mo iroiro, dasu hito mo,
And let's not forget the ones who receive them!"	Uketoru hito mo samazama ya.
"And now, in the Meiji era for a convenient fee	"Sate, benri no odai no kage,
Shikoku and Kyūshū are nothing—	Shikoku, Kyūshū nan no sono,
To the far north, as far as the Kuriles	Kita wa, Sento nosono wa hate made mo,
Or south, to the Ryūkyūs and even Taiwan:	Minami, Ryūkyū, Taiwan made mo
Parents, siblings, and friends, no need to travel,	Oyako, kyōdai, tomodachi dōshi,
News comes while they wait.	Tayori shira suru, inagara ni.
Above all, the familiar letters in a woman's hand	Wakete, natsukashi, kono onna moji.
'To Miss Oharu, a flower hairpin,	'Oharu-dono ni wa, hana-kanzashi o,
And to Master Tarō, a gym shirt	Tarō-dono ni wa, undō-shatsu o,

A gift from your aunt,' the
postcard reads
How children must feel: truly,
even to my eyes
the post box brings great
pleasure"[29]

Obagami miyage.' to kaitaru hagaki,

Kotachi, sazokashi, yūbinbako no

Ware ga mite sae, itodo tanoshi ya."

The differences in varieties of mail, handwritten addresses, and the people who send them may vary considerably, but through the flat rate for sending, the post box delivers the bounty from around the corner or the far-flung reaches of the empire to one's doorstep with equanimity. Such are the simple pleasures fit for children's nursery rhymes. But as was the case with the brisk military efficiency of the telegraph, post offices were indispensable to colonial and imperial ventures, laying the essential infrastructure of transportation and communication for all that followed. Japan's postal stakes in mainland Asia were established in 1876 with the establishment of a branch office in Shanghai. Although they were later arrivals than the Western concessions obtained after the Opium War in the 1840s, the Japanese would produce the most extensive postal network in Asia by 1920, by which time the Chinese at least temporarily succeeded in driving out the weakest of its foreign colonizers. They did so through military action and then by recourse to the Universal Postal Union (UPU), which from 1874 guaranteed equivalences in rates and fares and allowed nations to assume full sovereignty over the postal processes within their borders. John Mosher explains,

Up to the time of the first World War, China had no luck freeing herself of her postal rivals, but by declaring war on Austria and Germany (14 Mar. 1917) was able to evict those two countries. Russia followed in the wake of the Russian Revolution: recognition of Imperial Russia was rescinded 23 Sep. 1920. . . . Finally China entered the UPU and ordered England, France, Japan and the USA to close their post offices on 7 Jun. 1921.[30]

29. Shōyō Kyōkai, eds. *Shōyō senshū, bessatsu* 2:800–2. Translation is my own.
30. Mosher, *Japanese Post Offices in China and Manchuria*, xi.

This in no way contravenes what we have already noted about individual nations as the final arbiters of their participation in international conventions. Participation in the Universal Postal Union and the International Telegraph Convention in 1879 had also been among Japan's crucial steps in ridding itself of the unequal treaties, but war, or threat of war, remained the final recourse to asserting national sovereignty.

When called upon to survey the reach of the Japanese Imperial Post in its heyday circa 1910, Den Kenjirō, Japan's first minister of communications, and subsequently civilian governor of Taiwan in the 1920s, had much to report:

At present Japan possesses post-offices in Peking, Tientsin, Newchang, Chefoo, Shanghai, Nanking, Hankow, Shache, Soochow, Hangchow, Foochow and Amoy—twelve places in all—in China; also at Fusan, Wonsan, Masan, Mokpo, Kunsan, Chemulpo; Seoul, Pinyang, Chinnanpo, and Songjin, and other towns along the routes of the Seoul-Fusan and the Seoul-Ninsan railways. The Japanese postal service abroad is in every respect regulated in accordance with the postal laws in force at home, and except in a few cases, the rates of postage are the same as those obtaining within the empire.[31]

It goes without saying that this passage conveys only a very cursory sense of the complexities on the ground in the respective colonies. While Hokkaidō and Okinawa (1870) and the model colony of Taiwan (1895) were quickly assimilated into homeland discourse, treaty ports and occupied territories were relegated to a somewhat different set of rules. Although Mosher asserts that "from 1895 on Japan used every opportunity to establish new post offices in China and Manchuria" in its dominion over East Asia, he qualifies the difference between the scattershot approach to the former and the monopolistic situation in the latter:

The two cases developed differently. In China, Japan opened offices singly at specific locations usually named in treaties. In Manchuria they had a virtual open hand in the Kwantung Leased Territory and along the South

31.   Den, "Japanese Communications," in Ōkuma, ed, *Fifty Years in New Japan*, 1: 410.

2.3 *Heiwa kinen* (Peace memorial) postcard, 1919. Photograph by the author.

Manchurian Railway, which they saturated with installations of all kinds, including whole towns. Japanese postal activity in Manchuria was far more densely organized than in China proper.[32]

Despite the unevenness of extending postal service throughout the colonies, the post remained one of the key institutions for exercising and consolidating imperial control, bringing colonies closer to the rhythms of the nation through its homogenizing effects.

Den's final words therefore affirm the gist of Shōyō's "Post Box *Song*," namely, that the post not only links the colonies directly, but also permits equivalence to the *domestic* exchange value of letters and letter-writing. In short, the prestige of national and imperial belonging trumps economies

---

32. Mosher, *Japanese Post Offices in China and Manchuria*, xii.

of scale calculated by distance and volume. Moreover, from the moment postal treaties are concluded between sovereign nations, the domestic becomes interchangeable with the foreign in a scale of costs for sending and receiving. It then becomes the last reserve of counter stamps and customs officials to police and document the movement of paper trails across borders. A "Peace Memorial" post card (figure 2.3) featuring a young Japanese girl in a blue patterned kimono and a boy in his black, Prussian military-style school uniform surrounded by doves and dated 1919 reveals a simple truth in its payment instructions for postage stamps: "one *sen* five *rin* in the homeland, four *sen* abroad" (*naichi ni wa ichi-sen go-rin kitte, gaikoku ni wa yon-sen kitte*). Peace therefore does not come at any cost, but at a fixed cost set by nations and international conventions. In a fitting summation of the global link-up of national and colonial postal services at the end of the late nineteenth century, Siegert observes:

> The World Postal Union [UPU] transformed the concept of the post it-self: once global, uniform postage had obliterated the difference between homeland and foreign lands in the letter's form of being, the word "post" was essentially an abstraction from the various material carriers by which information was sent on the earth. . . . As the technical media them-selves finally were integrated slowly but surely into the postal system—in Germany, completely, in other states less so—"post" became synonymous with a General Medium for Transmission.[33]

The framing of the world, or at least its interconnectivity by a "General Medium for Transmission" would thus be chiefly conducted through the channels and codes of postal modernity.

33. Siegert, *Relays*, 144–45.

CHAPTER 3

Wiring Meiji Japan

From Hokusai's Postcard to Mokuami's Telegraph

The verbal, visual, and theatrical regimes of the urban townspeople commonly associated with the "floating world" (*ukiyo*)[1] were an integral part of Tokugawa popular culture that were neither firmly endorsed nor dependent upon the patronage of ruling elites. Nevertheless, it is from within these regimes in the first two decades of Meiji that the earliest iterations of the imagined community of the modern nation-state arose and in ways that often anticipated the broader discursive changes associated with national language and script reform to be discussed in part II, and the origins of modern Japanese literature in part III. This chapter contrasts representations of official and de facto communication networks bracketed by two texts at opposite ends of the epistemic rupture from late Edo to early Meiji: Katsushika Hokusai's woodblock print Shunshū Ejiri 駿州江尻 (Shunshū Station in Ejiri Province, c. 1830) from the collection *Thirty-Six Views of Mount Fuji* and the final act of Kawatake Mokuami's kabuki play *Shima Chidori Tsuki no Shiranami* 島千鳥月白波 (Plovers of the island and white waves of the moon, 1881), referred to hereafter as

---

1. Prior to the Edo period, *ukiyo* represented the Buddhist idea of a transient world of suffering. However, with the growth of towns and increase in commerce that followed the unification of the country under the Tokugawa shogunate in the seventeenth century, the entertainment districts peopled with courtesans, actors, and other disreputable types became new foci for artistic ideology and expression. While the aesthetics of evanescent sorrow by no means disappeared, they were reconfigured within the homophonic register of *ukiyo*, the "floating world" of fleeting bodily pleasures.

"The Thieves."[2] While Hokusai's print playfully captures the limits of writing in the sanctioned network of postal roads along the Tōkaidō, Mokuami's play propagandizes the new Meiji government's ideological reforms and technological innovations as being capable of transforming even the lowliest denizens of the underworld into loyal subjects.

The Postcard that Goes Undelivered

These discursive channels are manifested in a drawing by Matsumoto Shōun of the Sukiyabashi district located just outside the walls of Edo castle (figure 3.1), which appeared in the December 10, 1902, issue of the magazine *Fūzoku Gahō* 風俗画報 (Customs illustrated). Although its scale and figural conventions signal a debt to the woodblock print, it already represents an irrevocably altered landscape several decades into Meiji. Pack horses, rickshaws, canal skiffs, and pedestrians make up the commercial bustle along the main bridge and waterways, while laborers in Japanese dress and students in Western military-style uniforms gather around food stands advertising Westernized delicacies such as sweet red bean jam-filled pastries, and bowls of beef served over rice. What commands our attention, however, are the rows of thick, black telegraph cables suspended above the city streets. They are at once a visible reminder of the communications network taking shape around the nation and world, and of a technology concealed from the gaze, as electricity races across thousands of miles of black wire overhead.

Of course, this was not a place like any other. Located in the heart of Tokyo, the gateway into what would become the Imperial Palace between 1873 and 1888, it was circumscribed by the administrative and modern cultural institutions of the Meiji state. Just around the corner and

2. The title of the play can literally be translated as "Island Plovers, White Waves of the Moon." Donald Keene's translation of the title of this final act is serviceable, given that *shiranami* was an epithet for the popular genre of plays about thieves and the criminal underworld. Although in the original Japanese script the last act is simply labeled "act five" (*gomakume*), at least one contemporary literary historian, Tsubouchi Yūzō, refers to it as *Shōkonsha torii mae no ba* 招魂社鳥居前の場 (At the gates of Yasukuni Shrine).

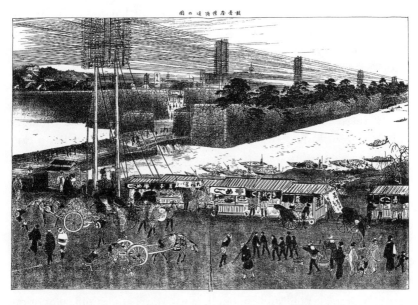

3.1 Matsumoto Shōun, *Sukiyabashi fukin* (In the vicinity of Sukiyabashi). Photograph courtesy of *Fūzoku Gahō* CD-ROM edition, published by Yumani Shōbō.

out of view from the picture plane of the scene depicted were the administrative offices of the Tokyo Metropolitan Government, the first Imperial Hotel, the Mitsubishi Bank, and major companies for postal shipping and electric lighting. Slightly farther off, on the other side of Hibiya Park, were both houses of the Diet, the offices of the Navy and Foreign Ministry, and the Russian, Italian, German, and British embassies.[3] The seat of power for nationalism, imperialism, capitalism, and industrialization was linked to the rest of the nation, and by extension the world, through the wires.

To be sure, before strands of telegraph and telephone wire darken the sky and postal systems encircle the globe, paper can still fly through the air and villains can vanish into thin air. The varying degrees of bureaucratic restriction on internal travel as a result of the shogunate's official "closed country" policy notwithstanding, the fifty-three stages of the

3. Tamai, ed., *Yomigaeru meiji no Tokyo*, 48–49.

Tōkaidō linked the shogun's capital in the east from Nihonbashi—symbolically the bridge from Edo to the rest of Japan—with the emperor's capital of Kyoto in the west. As such, the Tōkaidō comprised the main conduit between the population centers of the Kansai and Kantō plains for commerce, tourism, and religious pilgrimages. The Tōkaidō had been the official approach in the alternate-year-in-attendance (*sankin kōtai*) system for daimyo and their retinues coming from the south and west, as well as the retinue of Dutch traders in Nagasaki and tributary missions from Korea.[4] The ancient shrines at Mishima and Ise, meanwhile, were located in the vicinity of the twelfth and forty-forth stations from Nihonbashi, respectively. Besides the many famous places (*meisho*) of untrammeled beauty canonized in classical poetry, the way stations of the Tōkaidō were inscribed into the visual and verbal arts of Tokugawa woodblock print culture. The immense popularity of Jippensha Ikku's comic travelogue *Tōkaidō-chū Hizakurige* 東海道中膝栗毛 (Traveling the eastern seaway by Shank's mare), published in installments in kana booklets from 1802 until 1831, is often credited with boosting enthusiasm for travel. A veritable publishing boom in maps, travel guides and color prints ensued, including Hokusai's series *Thirty-Six Views of Mount Fuji.*[5]

For all their purported interest in travel, Katō Shūichi avers that townspeople such as Ikku and his ne'er-do-well protagonists scarcely acknowledged the existence of the provincial folk off the beaten path: "The heroes of the series . . . are Yajirōbē and Kitahachi, two low-class *chōnin* from the heart of Edo who travel along Japan's three major highways, the Tōkaidō, the Kisokaidō and the Nakasendō, encountering a variety of people and customs. They never leave the main road, however, or enter a farming village let alone come into contact with the inner world of a peasant family."[6]

While *Shank's Mare* undoubtedly contributed to a general sense of shared cultural topography, it was strictly demarcated along the lines of Tokugawa-era hierarchies and played up to the prejudices of an urban readership. Consequently as the heavily trafficked route for warrior

4. See, for instance, Ichige's "Chōsen Tsūshin-shi no Tōkaidō tsūkō," 229–68.

5. Despite its title, the series was comprised of forty-six, rather than thirty-six, prints. "Shunshū Ejiri" is the thirty-fifth in the series.

6. Katō, *A History of Japanese Literature*, 44.

retinues and travel-mad townspeople alike, the Tōkaidō was for all intents the late Tokugawa era's "trunk line" (*kansen*)[7] from which one did not stray while traveling.

Often missed in the discussions of natural beauty and local color along the way is the indispensable function of the Tōkaidō as a post road. The courier system established in 1663 and sponsored by the shogunate maintained two letter collection boxes (*shojō atsumebako*) at each stage of the Tōkaidō, one headed for Edo, the other for Kyoto. It was the hallmark of the relay-runners working in circuit to complete the distance from Edo to Kyoto in as little as thirty-six to forty-eight hours, depending on road conditions and inclement weather, compared to the usual ten to fourteen days for the average traveler. While private couriers also existed in the Three Cities and outlying domains, they operated on a considerably more modest and potentially insecure scale.[8] It was not only speed, but the reliability of the couriers that drew distinction to the way stations.

"Shunshū Station" (figure 3.2) takes its name from its location in Suruga Province, in today's Shizuoka Prefecture. Ejiri was the nineteenth of the fifty-three stages, renowned for the breathtaking beauty of Suruga Bay and Mount Fuji. It is almost exclusively in this context of representing natural landscape that the print has been evaluated by art historians in Japan as nothing more or less than a skillful, even formalistic, rendering of a blustery autumn day on the Tōkaidō highway.

Against the backdrop of these famous landmarks, travelers up and down the winding road duck down against the chill and clutch at their wide-brimmed straw hats. One hapless man has lost his hat to the wind, while a woman in a purple head covering in the foreground has let an entire sheaf of "tissue paper" (*futokorogami*)[9] be carried off by the wind. Two forlorn trees also in the foreground shed a scattering of leaves, which subtly flow toward a point of convergence with the paper somewhere on

7. Although the Shinkansen is known as the bullet train in English, it literally means "new trunk line."

8. See Maejima's "Communications in the Past," in Ōkuma, ed., *Fifty Years of New Japan*, 1: 396–401.

9. This "tissue paper" was not for blowing one's nose, but for composing songs (*uta*) and poems (*shi*). As the Chinese characters in its name suggest, it was paper kept close to one's heart in the inner breast pocket of a kimono.

3.2 Katsushika Hokusai, *Shunshū Ejiri* from *Thirty-Six Views of Mount Fuji*. Photograph © Trustees of the British Museum.

the horizon line just beyond the picture plane. Located in the center of the print below these flying objects, and very nearly camouflaged by the color and approximate size of the straw hats, is a small structure, likely a local shrine, which superficially resembles a letter collection box. That might prove a road too far, but how are we to make sense of the juxtaposition of paper and ink flying overhead?[10]

From the standpoint of these dual placements, "Shunshū Station" seems as self-consciously referential to the circulation and exchange of writing as it is one of Hokusai's innumerable paeans to the geometric perfection of Mount Fuji. Hokusai furnishes the landscape not only with elements of the natural and cultural sublime, but with fragments of the compositional and signifying process of the woodblock print—fragments that require us to reevaluate the print in terms of its surface meaning

10.   The post boxes of the Edo period were to be found only in the cities; villages and infrequently traveled roads such as this one—with no runners in sight—were not likely to have a collection box.

rather than as a "floating picture" imbued with the illusion of depth. The white sheets of paper rise through the air to meet a scattering of blue-black dots of ink like so many letters eager to be written. At the same time, this arrangement of objects lends itself to a kind of visual play that repeats and gathers itself in triangular patterns. The perfect isosceles triangle of Mount Fuji in the background is mirrored by the roof of the small structure in the center of the picture plane. Meanwhile, a third and less distinct triangulation is formed by the two tree trunks, the straight diagonal column of paper caught in the wind, and inkblot leaves heading toward it. These patterns are surely not coincidental. Considering that in premodern Japan leaves are tropes for both poems and words, and provide one of the two characters for the compound for "word" (*kotoba* 言葉), one might read the print according to its visual iconography as a play on the compositional process, a poem composed on the wind. References to leaves abound in classical and medieval texts such as the title of the eighth-century poetry anthology *Manyōshū* 万葉集 (Collection of myriad leaves) and the kana preface to the tenth-century *Kokin Wakashū* 古今和歌集 (Collection of Japanese songs new and old) by Ki no Tsurayuki, which begins: "The seeds of poetry lie in the human heart and grow into leaves of ten thousand [myriad] words."[11] The grass that blows in unison with the paper and leaves likewise calls to mind the semantic associations of the word for "gesture," which is sometimes written with the character for grass (*shigusa* 仕草, also 仕種). If these possibilities smack too heavily of European Romanticism—the gust of wind as an animating spirit or proto-voice; the cosmological convergence of natural patterns and man-made forms of writing—the leaves and paper that converge above and beyond the picture plane nevertheless pose an alternative circulation that *circumvents* the courier system. It is a scene of writing that solicits a "postal modernity" *avant la lettre*, or what in Japanese can properly be called Hokusai's post card.[12]

There is precedent for this apparent Romanticism in Motoori Norinaga's study of the literary aesthetics of *The Tale of Genji*, which centers on the concept of *mono no aware* (literally, "the pathos of things"). Ivan

---

11. Rodd and Henkenius, trans. *Kokinshū: A Collection of Poems Ancient and Modern*, 35.

12. The Japanese word for post card, *hagaki* 葉書, is composed of the characters for "leaf" and "writing."

Morris reminds us that *"Aware* never entirely lost its simple interjectional sense of Ah!" in Norinaga's recuperative philology of ancient Japanese speech: "the initial *a* of *aware* is cognate with the same sound that appears in other ancient exclamations like *haya, hamo* [and] *atsuhare.*"[13] It resonates with the contemporaneous notion of *Sprache* emanating from the primordial, phonocentric sigh *ach* that Kittler locates in Goethe's Faust and the discourse network of 1800 (that is, the German Romanticist canon). Political and romantic intrigue in the Heian court coupled and uncoupled as much through midnight letter exchanges as diplomatic alliances struck by day. It was a far cry from the anonymity of Hokusai's print and still further from the (fictions of) private, interiorized individuality we associate with the public channels of the late nineteenth-century postal service. As Morris writes:

> In the diaries and fictional works of the time there is a constant flow of letters and messages. Day and night, in fair weather and foul, the long-suffering messengers shuttle to and fro between the lordly mansions, carrying now a note on thin, white paper about a lady's emotions at the first snowfall, now a "next morning" letter, attached to a sprig of pine, in which a gentleman tells his mistress that his love will never wither. Since correspondence is regarded as an art form, there is little privacy about these letters; often they are read by the wrong people and this can result in endless complications.[14]

Further, as Thomas LaMarre has persuasively argued, we should not place all the value of classical poetry in its emotional expression at the expense of the calligraphic hand, the qualities of the fold and grain of the paper, and even the beauty of the messenger and gracefulness of delivery.[15] The Heian aesthetic was at once a secretive and yet intimately *public* performance, the conduct of affairs whose effects always ripple beyond the sheets of paper or beds.

Hokusai scholars such as Kondō Ichitarō and Nagata Seiji have interpreted "Shunshū Station" as a superlative instance of Hokusai's repre-

13. Morris, *The World of the Shining Prince*, 208. As Morris also notes, Motoori's studies of *mono no aware* bear a striking parallel with eighteenth-century European classicist studies of *lacrimae rerum*, "tears for things" (318–19).

14. Ibid., 200.

15. LaMarre, *Uncovering Heian Japan*, 98–100.

sentation of the effects of the wind as motion in the static medium of the woodblock print. It is a theme he devoted considerable attention to in his earlier workbooks called the *Manga*,[16] and there are a number of related drawings that they cite as earlier versions or variations of "Shunshū Station." Kondō points to the two figures and a winding road nearly identical to "Shunshū Station" in the sketch entitled *Sekiya no sato yū* 関屋の 里夕雨 (Evening shower at the village of Sekiya in Shimosa Province) in the *Manga*, volume 7 (figure 3.3).[17] Of course Hokusai had no intention of accurately representing the truth of the landscape in either picture. The winding road is a visual convention to guide the eye through the picture plane. Nagata, meanwhile, notes a detail in an untitled drawing in volume 12 (figure 3.4) of a woman vainly clinging to sheets of paper amidst a scattering of autumn maple leaves, which corresponds to the figure in the foreground of "Shunshū Station."[18] Along with this sketch, there is a drawing in volume 15 that depicts two travelers opening a chest outdoors on a windy autumn day only to find the paper inside carried off by the wind that can be added to this category (figure 3.5).[19] The latter two pictures do not have proper titles, but are simply marked with the single Chinese character for wind (風), almost invisible amongst the maple leaves the character is drawn to resemble. Whatever these three earlier precedents reveal about Hokusai's reliance on sketches and models from the *Manga*, it does not address the connections of wind, paper, and leaves. Wind is a powerful trope for signification: an animating, yet invisible presence. Consequently, representing the wind visually, with its mobilization of paper and leaves to form compositions, is a case of substitution par excellence.

Another sketch from the *Manga* suggests the wind as an elemental force of poetic composition. "Chōkyūka" 張九歌 (Songs of Zhang Gui)

16. Although "manga" nowadays is used to designate a contemporary Japanese genre and industry distinct from Western (usually American) comic books, the term originally meant much the same thing as "cartoon" does in English or *bandes dessinées* in French: a series of preparatory sketches or drawings prior to a more complex painting or sculpture.

17. Kondō, ed., *The Thirty-Six Views of Mount Fuji by Hokusai*, plate 35. The image is reprinted in Nagata, ed., *Hokusai Manga*, 2:76.

18. Nagata, *Fūkeiga 2*, 58. The image is reprinted in Nagata, ed., *Hokusai Manga*, 3:84–85.

19. Nagata, ed., *Hokusai Manga*, 3:285.

3.3 Katsushika Hokusai, *Shimosa Sekiya sato no yūdachi* (Evening shower at the village of Sekiya in Shimosa) from the *Manga*. Photograph courtesy of Yale University Library.

3.4 Katsushika Hokusai, *"Kaze"* (wind) detail from the *Manga*. Photograph courtesy of Yale University Library.

3.5 Katsushika Hokusai, Detail from untitled drawing from the *Manga*. Photograph
courtesy of Yale University Library.

depicts the Chinese statesman-poet Zhang Gui (301–376 CE) making
butterfly-shaped paper cutouts with a pair of scissors, which fly from
his open hand seemingly of their own free will (figure 3.6).[20] While the
butterfly invariably calls to mind Chuang Tzu's famous dream of the
butterfly,[21] it also suggests the sophisticated play of Edo poetry-prints
such as *surimono*, whose circulation and exchange were integral to the
social order of the floating world.

It is important to keep in mind that the floating pictures executed
in woodblock arose from a longstanding dialog with the paintings,
copperplate etchings, and perspectival drawing introduced by Dutch

20. Ibid., 295.
21. The enigmatic passage from the *Chuang Tzu* runs as follows:

> Once upon a time Chuang Chou dreamed that he was a butterfly, a butterfly flit-
> ting about happily enjoying himself. He didn't know he was Chou. Suddenly he
> awoke and was palpably Chou. He did not know whether he was Chou who had
> dreamed of being a butterfly or a butterfly dreaming he was Chou. Now, there must
> be a difference between Chou and the butterfly. This is called the transformation
> of things. (Mair, *Wandering on the Way*, 24)

3.6 Katsushika Hokusai, "Chōkyūka" from the *Manga*. Photograph courtesy of Yale University Library.

Learning.[22] Beginning with Shiba Kōkan, Aōdō Denzen, and others who experimented with these techniques, the popular prints available at lending libraries and print shops were indelibly associated with a vogue for optical and mimetic games. Unlike the generic figures and mistenshrouded, idealized landscapes of the Kano and Tosa schools, floating pictures increasingly sought to represent contemporaneity through a variety of literal and figural lenses. Their technologies of the gaze express a worldview often at odds with the Neo-Confucian ideals sanctioned by the government: the science of Dutch Learning coupled with the sensibilities, aspirations, and indulgences of an increasingly wealthy, but politically disenfranchised, merchant class.

Unlike the mass dissemination of knowledge under the ethos of Civilization and Enlightenment in Meiji, the materials that came in sporadic intervals from Holland to the small settlement in Dejima and were translated by Dutch Learning scholars have often been understood as amusements for sophisticated urbanites or contemplation by a learned few. Noting the widespread misconception that the pleasure quarters of major cities and towns where Dutch Learning texts thrived were simply sites of apolitical carnal indulgence, Timon Screech insists they offered a measure of intellectual freedom outside the bounds of official ideology or conduct:

> The mental spaces of leisure activity are often billed as mere drollery, but they are more than that: beyond flight from life, or ludic reenactments of it, the Edo sense of play (*asobi*) was a full rescripting of experience. Ideas and manners not formulated by government dictate spun at the peripheries like orbs, but with gravitational pulls of their own. Neither revolutionary nor fully accepting, these lay athwart the division between contestation and compensation, unable to impinge on the centre, nor yet quite distant from it.[23]

Beyond the immediate compass of the floating world, publishers, private academies, and other institutions outside the government were obliged

22. Dutch Learning is best understood as a school of learning that took place in the absence or limited presence of European persons; knowledge was obtained predominantly through translation and interpretation of texts, art objects, and scientific instruments brought to Japan.

23. Screech, *The Lens within the Heart*, 23.

to preemptively censor themselves. Jail, exile, and seizure of one's property were all punishments that awaited those who courted controversy by representing issues that called into question the right or righteousness of Tokugawa rule. Yet it is also true that while the shogunate had an elaborate spy system and many layers of bureaucracy to monitor activities throughout the realm, markets and institutions frequently went unregulated or were inconsistently regulated. Without an edict or law in place explicitly prohibiting a certain product or form of knowledge—and indeed, the fields and texts of Dutch Learning were generally permissible so long as they did not make reference to Christianity—intellectuals operated in a gray area, the sort that led to circulation of ideas amongst colleagues without publishing them outright. In such a climate of semiprohibition, Hokusai's print may be seen to hide in plain sight an astute political commentary that would not have been permitted had it been stated more forthrightly.

The shogunate continued to demonstrate for the next several decades what must seem to us now an almost incredible refusal to grasp the implications of the technological transformations that led to an embryonic imagined community, especially through the medium of vernacular print languages. The use of print or post as subjective technologies for consolidating state power were almost unheard of until the last decades of its rule, and as in the case of Maejima's call for an end to Chinese characters, they went unheeded. Instead, censorship was used to curb access to ideologically incendiary materials and as a bulwark against other forms of social protest. These policies worked domestically to the advantage of the shogunate and provincial daimyo, who did little to rectify the linguistic heterogeneity in the realm or inculcate widespread patriotism on the order of modern nationalism. They were also ineffective at keeping out foreign interventions by the early 1850s.

On the other hand, when sent abroad on one of the early fact-finding missions of the shogunate in 1860, a young Fukuzawa Yukichi was quick to realize the tremendous potential of modern media. Recounting his firsthand glimpse of the power of the printing presses in the essay "London Newspapers," he observed:

> The aim of newspaper reports is speed, and by means of the steam engine for printing, it can produce 15,000 pages an hour. When the binding is

done, it is promptly delivered to all the necessary places by steam-driven locomotives and steamboats. This rapidity arrests people's attention. . . . It is of great value to the people when there is great debate: it can at once influence their hearts and it can also be used for reform in discussions of government.[24]

Within the more restricted scope of the woodblock print, avoiding any direct impingement upon proscribed topics allowed for alternative, gray-market perspectives and ideologies to be registered. Beyond the intentionality that can be attributed to Hokusai, "Shunshū Station" may at least retrospectively be seen to celebrate the opportunity to skirt past the channels of official discourse networks with its convergence of paper and leaves of words just beyond the picture plane. The natural and cultural imagery in "Shunshū Station" therefore cannot simply be called national landscape, nor can Hokusai's mobilization of poetic tropes and post be reduced to an act of national writing. However much it subverts official channels and codes, Hokusai's play on poetic words and images resists enlistment into state service and stops short of fulfilling any promise of national community[25]—it is the post card that goes undelivered.

Telegraphing the Imagined Community from Yasukuni Shrine

In stark contrast to Hokusai's free-spirited prints where leaves of words can still escape the system of censors and police, Mokuami represents the looming panopticism of the state in the spate of early Meiji kabuki plays epitomized by "The Thieves." Yet before advancing to this play, it is worth pausing for a moment to consider another woodblock print from the Meiji period, one whose discourse networks are also laid at the

---

24. Iida, *Nihon ni okeru kagaku gijutsu shisō no keisei*, 465. The translation is my own.

25. It is not hard to imagine a different sort of image altogether—the red rays of the rising sun shining through the center of white pages—to spread the message of imperial nationalism. Kawabata Yasunari did as much with a snow-covered landscape in *Snow Country* written at the height of the Pacific War.

3.7 Kobayashi Kiyochika's *Hakone sanchū yori Fugaku chōbō* (Distant view of Mount Fuji from the Hakone Mountains), 1877. Photograph © 2015 Museum of Fine Arts, Boston.

foot of Mount Fuji. If isomorphic comparison in the annals of art history were our only purpose, we could find no better example of the epistemic rupture than to look at Kiyochika's "Hakone sanchū yori Fugaku chōbō ichigatsu jojun gogo sanji utsushi" 従箱根山中富嶽眺望 一月上旬午後三時写 (Distant view of Mount Fuji from the Hakone Mountains—Sketched at 3 p.m. in the afternoon in late January 1877; figure 3.7). As the verbose title written in the manner of an Impressionist reflects, Kiyochika recorded the specific coordinates of clock and calendar as well as place that factored into its composition. In contrast to Hokusai's idealized brocade print that was executed in a collective manner, Kiyochika further distinguished his break with Tokugawa traditions by designing prints by himself. The grounds of landscape are forever changed, too. Kiyochika's Fuji has no force: it is a hazy outline in the background, a ghost of the majestic cone worshipped by Hokusai. Instead it recedes behind the realistically depicted foothills that partially cover its girth. Telegraph lines parallel to the cobblestone road stretch high above Fuji. It is a striking statement about the already complete transformation of the countryside into a national space where Fuji is subordinate to the landscape of nation and media. As we shall see, the hallowed ground of the common people has also shifted elsewhere to a site of the state's creation.

One is reminded of the devastating words of wisdom delivered in Natsume Sōseki's *Sanshirō* (1908) by the enigmatic stranger on the train to the impressionable young man heading to Tokyo for the first time from rural Kumamoto: "You've never seen Mount Fuji. We go by it a little farther on. Have a look. It's the finest thing Japan has to offer, the only thing we have to boast about. The trouble is, of course, it's just a natural object. It's been sitting there for all time. We didn't make it."[26] In Kiyochika's similarly dispassionate reappraisal of the iconic mountain, Fuji and the golden era of the brocade print all but disappear from view.

In a similar regard, post-Restoration plays have long been criticized for violating longstanding conventions of kabuki and positively heralding the new age of "Enlightened Civilization." For conservative Meiji audiences and postwar literary scholars alike, it was less an acknowledgment of events following the Meiji Restoration that were deemed objectionable than the supposedly inauthentic mode of realism that characterized their portrayal. I contend, however, that "The Thieves" not only provides disquieting insights into compulsory nationalism by foregrounding the wiring of the imagined community by modern postal and print technology, but that its uses of realism were instrumental to effecting that epistemic break.

The intensification of censorship under the Kyōbushō (Ministry of Religious Education) during its brief tenure from 1872 to 1877, as well as ongoing regulations of theatrical performances and prohibitions against criticism of the state into the 1880s, unquestionably contributed to early Meiji kabuki's ubiquitous formula of *kanzen chōaku*, or "punish evil, reward good," in the service of the nation-state. Yet neither can we overlook the efforts of Mokuami and leading actors such as Danjuro IX (1838–1903) and Onoe Kikugorō V (1844–1903) to move kabuki away from its allegorical strategies of representing the present in the past. Instead, they

26. Sōseki, *Sanshiro*, 15–16.

set their works in the Meiji era and injected a new mode of realism into their performances by speaking in vernacular speech, wearing modern clothes, and so on. Barred from expressing political views or even commenting on ideologically controversial subjects, playwrights previously overcame these obstacles by working within two genres, the *sewa-mono*, or plays dealing with contemporary situations such as romances or vendettas between commoners, and *jidai-mono*, or period pieces. Particularly in the latter, elaborate systems of indirect reference literally called "worlds" (*sekai*) were built up and sought to work around these prohibitions. Yet Mokuami's plays written on behalf of, or in collaboration with, these actors cannot simply be understood in terms of a facile endorsement or opposition to state power. Rather, these are works in which new print media and postal technologies (such as the newspaper and telegraph in "The Thieves") mark the tensions between the ideals of national ideology and the ethos of the floating world.

At a time when kabuki was slowly being compromised by competing political and aesthetic forces, Danjurō IX's attempts at representing contemporaneous events in the 1870s were mockingly labeled *katsureki-shi*, or "living history," by one of the last of the gesaku writers, Kanagaki Robun. Similarly, Mokuami's *zangirimono*, or makeup-less, "shorn hair" plays for Onoe Kikugorō V often drew upon newspaper reports and other modern sources of inspiration. Otherwise retaining the conservative tendencies of kabuki (all male ensembles, traditional acting techniques, and so on), both of these versions of modern kabuki were poorly received by contemporary audiences. These setbacks, coupled with the negative reassessment of kabuki as "old-style" (*kyūha*) theater versus the "new style" (*shinpa*) somewhat more closely adapted from Western theater, contributed to a sense of crisis in the theater and a fear of venturing too far from established norms. Further compounded by the defection of low-ranking actors and understudies, who had chafed under kabuki's internal hierarchies, to the silent screen in the 1910s and 1920s, Meiji kabuki has been retrospectively marked as the beginning of a long, slow decline from its heyday in the Edo period. *Katsurekishi* and *zangirimono*, which were received with mixed critical and popular success, have been overwhelmingly reconfigured in postwar scholarship as failed experiments. However, to dismiss them in this fashion is to overlook not only their contributions to the discourse of realism, which gave rise to modern theater, fine art, and literature, but also their performance of a compulsory nationalism solicited by new relations of media and language (as well as government intervention) in the 1870s and 1880s.[27]

Despite Robun's sarcasm, *katsurekishi* and *zangirimono* would be reevaluated in a more favorable light by later reformers of theater and literature, notably Tsubouchi Shōyō, for whom these set a critical precedent for the future of realist literature. We will return to this issue in the discussion in chapter 8 of Shōyō's *Shōsetsu Shinzui* 小説真髄 (Essence of the novel, 1885) and the essay *"Bi to wa nan zo ya"* (What is beauty?, 1886), where Shōyō calls for a balance between traditional and modern modes of representation. At the same time, we should also note the uncanny resonance of *katsurekishi* with *katsuji* (movable type) and *katsudō shashin* (moving pictures), technologies that, by the end of the century, would decisively contribute to the foreclosure of kabuki and the shift toward maintaining the imagined community by other technical means. These events are inimitably woven into the narrative fabric of Mokuami's plays.

"The Thieves" is organized around the reunion of the yakuza blood brothers Matsushima Senta and Akashi Shimazō. They meet under the cover of darkness to discuss Senta's plans for revenge: the vendetta is against the moneylender Mochizuki, who married a geisha Senta fancied and nearly ruined himself financially by pursuing. Yet contrary to Senta's expectations, Shimazō is reluctant to join him, protesting instead that he has had a change of heart and decided to reform his ways. What follows is a confrontation between two diametrically opposite modes of subjectivity. Even as Senta calls upon Shimazō to honor the intimately homosocial bond that united them as blood brothers (a pact they made together in prison), Shimazō appeals to the higher authority and nobility of sacrifice of modern nationalism.

Despite their criminal status at the outset, Senta and Shimazō's relationship aptly expresses the idealized conditions of the tightly knit

27. A useful overview of the source material is provided in Tschudin's "Danjurō's *katsureki-geki* and the Meiji 'Theatre Reform' Movement." I disagree with his conclusions, however, that when "carried to its logical end, katsureki could only lead to a dried-up and austere form, deprived of the greater part of kabuki's seductive power" (*Japan Forum* 11: 1 [1999]: 91). As we will also see in chapter 8, Shōyō and other leading critics were not necessarily opposed to these realist experiments in the kabuki theater.

premodern community.[28] Comparing the transformed relations of political and lived identity in the "imagined community" of the modern nation-state theorized by Benedict Anderson to such earlier forms of governance, Eric Hobsbawm observes,

> The "homeland" was the locus of a *real* community of human beings with real social relations with each other, not the imaginary community which creates some sort of bond between members of a population of tens—today even of hundreds—of millions. Vocabulary itself proves this. In Spanish *patria* did not become coterminous with Spain until late in the nineteenth century. In the eighteenth century it still meant simply the place or town where a person was born. *Paese* in Italian ("country") and *pueblo* in Spanish ("people") can and do still mean a village as well as the national territory or its inhabitants. Nationalism and the state took over the associations of kin, neighbours and home ground, for territories and populations of a size and scale that turned them into metaphors."[29]

Although we must be careful not to be drawn into uncritical usage of weighted language such as "real" and "imaginary," which unintentionally resonates with the tripartite structure of Lacanian psychoanalysis, Hobsbawm's position can readily be understood in terms of a widening gulf between presence and representation. That is to say, he calls attention to the intimacy of a social relation that is at once also political and familial, signifying connections to blood and soil that held sway in small, primarily agricultural, localities. To what extent this is an accurate description of *macroscopic* premodern social hierarchies, the principalities and kingdoms that ruled over the villages and countrysides, remains another question. As Anderson points out, unlike the modern nation state that strives to form a horizontal relationship amongst its citizens through the use of a common national language, there was little concern in premodern times about linguistic or cultural difference between ruler and ruled.

Finer distinctions aside, Hobsbawm's remarks about the semantic metamorphosis of country and people also hold true for the term *kuni*,

---

28. These are arguably the same romanticized notions of family, tribal, and clan loyalties that continue to inspire devotion to the genre of yakuza and Mafia films today.
29. Hobsbawm, *The Age of Empire: 1875–1914*, 148.

or "country," in Japanese. While *kuni* traditionally designated the birth-place or region from which one hailed,[30] in the aftermath of the Meiji Restoration, it became the basis for nation as both a people and state. Indeed, as Mark Ravina explains, the Meiji government embarked on a policy of eradicating the "feudal" terminology in order to solidify the sense of Japan as the sole legitimate basis for the nation as one's place of origin:

> Under the pretext of restoring seventh- and eighth-century political insti-tutions, the Meiji government eliminated the word kuni as a term for do-main. Implicitly, Japan became the only effective country/kuni, the Meiji state the sole state/kuni, and the Japanese people the only people/kokumin. The introduction of distinct terms for prefecture, domain and state was thus part of the construction of the modern state itself.[31]

It is this semantic difference between *one's people* and *the people* that mo-tivates the tension between Senta and Shimazo's divided and incompat-ible loyalties.

It is all the more fitting, then, that the setting for "The Thieves" is the Yasukuni Shrine, established in 1879 as a place of repose for the spir-its of those who died fighting for the emperor.[32] As Anderson argues, the tomb of the unknown soldier, itself unknown to premodern religious and

---

30. By the *kokugunrisei* (国郡里制) and more ancient systems of land apportionment, *kuni* was originally a unit of measure designating the largest seat of political power.

31. Ravina, *Land and Lordship in Early Modern Japan*, 33.

32. In the eyes of many in Japan's former colonies and in left-wing and pacificist circles today, the Yasukuni Shrine inters nothing so much as the bitter memory of Japa-nese imperial aggression and war crimes. Meanwhile, the officials overseeing the shrine continue to indulge in fantasies of Japanese righteousness in the Pacific War, as evi-denced by their bald-faced attempts at propaganda to schoolchildren in the *Yasukuni daihyakka: watashitachi no yasukuni jinja*. As it explains, amongst the 2.5 million *kami* (spirits/gods) contained by the shrine,

> There are also those here who took the responsibility for the war upon themselves and ended their own lives when the Great Pacific War ended. There are also 1,068 who had their lives cruelly taken after the war when they were falsely and one-sidedly branded as "war criminals" by the kangaroo court of the Allies who had fought Japan. At Yasukuni Shrine we call these the "Shōwa Martyrs" [including General Tōjō Hideki], and they are all worshipped as gods. (671)

dynastic societies, is a monument unique to the modern nation that sacralizes the quasi-religious function of nationalism regarding issues of mortality and faith into the "secular transformation of fatality into continuity, contingency into meaning."[33] For Mokuami, the shrine becomes the literal backdrop for the assimilation of the blood brotherhood of the underworld of crime into the horizontal fraternity of nationalism.

The vendetta as a staple of Tokugawa theater is also fatally undermined by this play, which establishes the law of the state as having the final word on all matters of life and death. In *The Structure of World History*, Karatani argues that the vendetta is a mode of reciprocity practiced by primitive societies, which is antithetical to the modern nation-state:

> When a member of one community is murdered by a member of another community, revenge (reciprocation) is pursued. The "obligation" for reciprocation here strongly resembles the "obligation" of gift-countergift. . . . But once a vendetta is initiated and revenge obtained, this in turn must be reciprocated, so that the process continues without end. . . . Vendetta is abolished only when a higher-order structure capable of sitting in judgment of crime arises: the state. This shows, in reverse, how the existence of vendetta impedes the formation of a state. This is because vendetta restores the independence of each community from the higher-order structure.[34]

Senta's desire for revenge must not only be countered, but overcome by Shimazō's insistence upon the benevolence and rectitude of the state. With some significant exceptions such as Ogyu Sorai's commentary on the revenge-killing of Kira Yoshinaka by the retainers of Asano Naganori—the events immortalized in *Chushingura* (The treasury of loyal retainers)—the shogunate displayed a remarkably tolerant attitude toward the vicious cycles of violence glorified in the townspeople's theaters.

Accordingly, "The Thieves" emphasizes the enlightened turn in Meiji resulting in the reform of the Tokugawa regime's exercises of discipline and punishment. Shimazō attempts to persuade Senta that the compassion of the emperor will not only spare them the death penalty if they surrender to the authorities and return the money they have stolen, but

33. Anderson, *Imagined Communities*, 11.
34. Karatani, *The Structure of World History*, 41.

commute a reduced sentence of ten years of hard labor to seven or perhaps even five years. Equally disquieting is how this penal system is yoked
to a notion of rehabilitation intended to change the habits and conduct
of the wrongdoer, as Shimazō proselytizes to Senta: "Would you rather
be praised and enabled to live out your full life, and even be of some
service to your country, or abused and forced to die?" (*isasaka ue no go-
hōkōnin ni homerarete ikinobiru ka, waruku iwarete inochi wo suteru ka*).[35]
Conversely, by submitting to the punishment of hard labor, they will be
allowed to expiate their debt to society and eventually reenter society on
equal terms with other citizens.

> SHIMAZŌ: And when once you serve out the sentence all the crimes you
> have committed are washed into the sea and completely forgotten. If you
> can really pass this test and work day and night with all your strength.
>
> NARRATOR: It is certain you will receive Blessings from Heaven which
> once punished you.[36]

There is nothing less than a religious conversion at work here. Their previous crimes and Senta's vendetta plot are transmuted into atonement and
a newfound sense of mission to the nation—the nation becoming the
transcendental Other to whose higher order Shimazō, Senta, and even
the moneylender, Mochizuki, are reconciled. We might also note that
while the play's economy of rights balanced with wrongs and settlement
of moral debts with money is superficially coded as Buddhist, it ultimately
affirms the true path of nationalism. This is readily evident when, just as
Senta and Shimazō agree to forgo the vendetta if Senta can get enough
money to settle his debts, a deus ex machina intervenes in the form of Mochizuki leaping from the bushes to admit his own faults and settle their
mutual debts. "Pardoned in the general amnesty after the Restoration"
(*go-ishin go ni tenchō kara hijō no dai shade hōmen sare*),[37] Mochizuki,
too, has had a change of heart since the days when he was an influential
retainer of the shogunate who abused his position to harass and intimidate

35. Keene, ed., *Modern Japanese Literature*, 48. In Japanese, see *Mokuami Kyakkuhonshū*
(hereafter *MKS*), 12:617.

36. Ibid., 45. *MKS*, 2:613.

37. Ibid., 41. *MSK*, 2:607.

commoners. Admitting to Senta and Shimazō that he was once no better than they, he promises to help them make amends, assisting in their return to the service of the almost as recently "restored" emperor.

The nation is thus organized as a programmatic set of rules, practices, and codes—in short, as a regime—which carries out its authority through the diversification and intensification of discipline. Foucault articulates this development in his essay "The Subject and Power," providing additional clarification of the principles of panoptic state control expressed in *Discipline and Punish*:

> What is to be understood by the disciplining of societies in Europe since the eighteenth century is not, of course, that the individuals who are part of them become more and more obedient, nor that all societies become like barracks, schools, or prisons; rather, it is that an increasingly controlled, more rational, and economic process of adjustment has been sought between productive activities, communications networks, and the play of power relations.[38]

This is the logic of nationalism at work, cultivating from within the desire for citizenship in the modern Meiji state by the three former recalcitrants. It is not coercion by force that motivates their actions, but the adjustment of forces, including the unseen power of karma, that motivates their identification with the nation.

It is also the application of disciplinary apparatuses and technologies. Perhaps the best known put to use in the service of the modern nation-state is forensic photography, which was used to create the first standardized criminal records instrument in mid-nineteenth-century Europe and the Americas. Alan Sekula and Tom Gunning have described how the earliest uses of photography beginning with the mug shot—so named for the contorted faces the first generations of clever criminals used to thwart the camera from accurately recording their faces—became the most formidable tool in the budding science of criminology. Although photography had already been in use to document prisoners since the 1850s, French police statistician Alphonse Bertillon instituted a new and more rigorous application of "metric photography" to create a database of criminals. As Gunning explains,

38. Faubton, ed., *Power*, 339.

First, Bertillon systematized the process of police photography. He standardized the distance of subject from camera; created a special chair on which the subject would sit and which would control physical position and posture; determined the type of lens, thereby introducing a closer and unvarying framing; and established the directly frontal and profile angles of the now-familiar mug shot. These procedures gave criminal photography a consistency that facilitated its use as information and evidence. Further, it established the process of photography as a disciplinary process, asserting the system's power over the criminal's body and image. The system determined the look and posture within the photograph; the criminal simply delivered up the facticity of his or her body.[39]

What is determined by this use of photography, in other words, is its capacity to extend the surveillance of the state and policing of its borders across physical distances and human populations. It is in the same vein that Foucault points to the architectural blueprint of Bentham's Panopticon as a primarily *visual* technological apparatus that internalizes constant self-surveillance upon those who are confined within in midst.

In a similar respect, Mokuami's "The Thieves" connects the encompassing of territory and enforcement of state power to newspaper and telegraph, two media that effectively demarcate the interior of the national community. As with Bertillon's photography, being branded an enemy of the state begins with entry of one's name into the records. Nor does it end there, because with each new invention for sending and receiving information, we are woven tighter into the fabric of mass-mediated social relations. Kittler has remarked that the true horrors of Bram Stoker's *Dracula* (1897) are not to be found in the figure of the blood-sucking, decadent Old World aristocrat, but in the fast-tightening webs of bureaucratic red tape—the paper trails that disclose the location of his coffins—which lead to his destruction. Hence there is no escape for a villain on the kabuki stage in a nation whose provinces and population

39. Gunning, "Tracing the Individual Body," in Charney and Schwartz, eds., *Cinema and the Invention of Modern Life*, 29–30. Needless to say, Bertillon's previous system of anthropomorphic measurements of the human body only further underscores the degree to which standardization was put in the employ of state power. In time joined by fingerprinting and forensic science, Bertillonage and the mug shot force the body to yield its indexical truths regardless of what the subject, or accused, has to say in the matter.

are encircled by modern communication technologies. Not even the spaces off the beaten path of the Tōkaidō envisioned by Hokusai can any longer guarantee successful evasion from the tightening web, as Shimazō warns Senta: "An alarm will be telegraphed all over the country. You'll never escape. You'll be caught in three days" (*denshin de shirase ga mawareba nigerarenē, mikka to tatazu toraerare*).[40]

Mokuami's earlier *zangirimono* play, *Tokyo Nichinichi Shimbun* 東京日々新聞 (Tokyo Daily Newspaper, 1873), offers an astonishingly similar variation on these themes of new media, identification with nationalism at all costs, and reflections upon the "wiring" of the nation-state. The typically convoluted web of relationships unfolds as follows:

> The central character, Torigoe Jinnai, is a rōnin [masterless samurai] who has led a dissolute life ever since the changes brought about by the Meiji government ended his usefulness as a samurai. One day the prosperous townsman Chichibuya Hanzaemon, a man well known for his charitable deeds, accidentally overhears a young couple who are planning to kill themselves because of their desperate circumstances. He gives them seventy yen to surmount the crisis. After the young people happily depart, Jinnai drunkenly stumbles onto the scene, and for no good reason kills Hanzaemon. The young couple are traced by the serial numbers on the bills given them by their benefactor, and are arrested on a charge of having killed him. Jinnai learns of their predicament from the Tokyo Nichinichi Shimbun and, moved to contrition, sends a telegram to the authorities confessing his guilt. After killing Hanzaemon he had fled to Kobe, but now intends to surrender to the Tokyo police as soon as possible. He tells an acquaintance: "Fortunately I have the money I raised by selling my swords. If I take a steamship from Kobe, I'll be in Yokohama in three days. I'll take a train as soon as I arrive, and on the same day I'll give myself up in Tokyo and accept the punishment I deserve for having killed a man."[41]

It is media capture in more ways than one. Not only does the title of the play borrow its name from the newspaper whose article inspired the story in the first place, but there are the same expressions of interconnected networks of communication and transportation networks as "The Thieves"

40. Keene, ed., *Modern Japanese Literature*, 45.
41. Keene, *Dawn to the West*, 2:404.

that solicit national belonging. The immediacy of the telegram to stave off punishing the innocent and provide immediate gratification to the eagerly repentant Jinnai is further reinforced by his use of steamboat and rail (the latter newly built, and the only line in the nation) so as not to delay the swift meting out of justice. While the desire to be caught and punished may strike some audiences as dubious human behavior, as we see in "The Thieves," the tables could easily be turned to ensnare those resisting arrest. Hence the more salient issue is how it attests to the shrinking of national time and space, and to the proliferation of traceable standardized measures such as the serial numbers on the paper currency that lead (mistakenly) to the arrest of the young couple.[42]

Mokuami's self-conscious reflections upon the fading information value of kabuki in "The Thieves" are further epitomized by the figure of a roadside noodle seller who, at the opening curtain, informs two pilgrims of the significance of the shrine. The noodle seller metonymically serves as a figure for kabuki itself. The dialog between the two pilgrims and the noodle seller runs to the effect that while his noodles are delicious and no doubt valued by the residents of the neighborhood, the trade he plies grows increasingly scarce in the modernizing city:

FIRST PILGRIM: There seem to be fewer people selling noodles at night nowadays.

NOODLE SELLER: That's right. They're all in the suburbs [outlying Yamanote area], and nobody's left downtown.[43]

In the self-reflexive exchange that follows, one of the pilgrims asks the noodle seller the time and place of their whereabouts, prompting the other

42. In the West there was already ample precedent for the telegraph being used as a noose tightening around the necks of criminals. Tom Standage relates several examples from Victorian England in which pickpockets in Paddington Station who used the train as their getaway car were apprehended by police using the telegraph to coordinate their dragnet. Likewise, the telegraph was instrumental in the capture of the murderer John Tawell on the Paddington-Slough Line on January 3, 1845. "Tawell was subsequently convicted and hanged, and the telegraph gained further notoriety as 'the cords that hung John Tawell'" (*The Victorian Internet*, 50–51). It is entirely possible that Mokuami drew upon news reports such as this disseminated in the newspaper wire service reports to create this final scene.

43. Keene, ed., *Modern Japanese Literature*, 38. *MSK*, 2:603.

to remark, "That's like a line out of a kabuki play, isn't it—asking the noodle seller the time" (*yoku shibai de mo suru koto da ga, sobaya-san to iu to jikan wo kiku ne*).[44] It is a comic touch that softens the lament for the waning influence of the kabuki. That loss of influence strikingly contrasts with the power of new postal and print media euphorically proclaimed by Senta upon his decision to join Shimazo and Mochizuki as a new citizen of the nation at the conclusion of the act: "If word of this appeared in the newspaper it would set a good example for thieves" (*kono kotogara ga shimbun e detaraba zoku no yoi oshie*).[45] If Hokusai's postcard remains nothing more than a possibility just over the horizon, Mokuami's telegraph delivers its nationalist message with alacrity.

Kabuki had long been a repository of visual spectacles, adapting dramatic techniques from noh and bunraku on the one hand, and establishing a host of stylized gestures and scenes captured in woodblock print culture through genres such as the actor print, or *yakusha-e*, on the other. The hegemony of the late Edo regimes of representation maintained by the theater and woodblock print culture was sorely compromised by the rise of new print media and the wiring of Meiji Japan. As Katō Shūichi notes, Mokuami was the consummate, and ultimate, kabuki playwright before its eclipse. Mokuami is

> generally regarded as the last notable Kabuki playwright [who] wrote pieces which amount virtually to a sum of all the *Kabuki* stage-effects. In his most typical pieces—known as *shiranami-mono* (robber pieces), in which the hero is an outlaw—the dramatic opposition between duty and emotion that is at the centre of Chikamatsu's puppet theatre is supplanted by a world where duty and emotion coexist amidst a whirl of scene changes using a revolving stage, quick costume changes (a technique admired for itself), exaggerated and stylized poses and speeches of colourfully and ringingly delivered abuse and defiance in the Edo dialect.[46]

44. Keene, ed., *Modern Japanese Literature*, 39. *MSK*, 2:605.
45. Ibid., 51. *MSK*, 2:623.
46. Katō, *A History of Japanese Literature*, 11. It should be noted that *Shima chidori tsuki no shiranami* represents a significant break with the apolitical theatrical spectacles of Mokuami's pre-Restoration works such as *Sannin kichisa, kuruwa no hatsugai* (The three Kichisas: first night in the pleasure quarter) from 1860.

Whereas kabuki had been at the center of life in the pleasure quarters, it was irrevocably diminished by reforms in Meiji, ironically including those events meant to confer greater legitimacy such as the Meiji emperor's attendance at a kabuki performance in 1887. Previously unthinkable in the centuries when kabuki actors were considered "riverbank beggars," the imperial visit presaged the decline of kabuki from the pinnacle of urban merchant culture into the more rarefied context of a national theatrical art. By the same measure, as Maeda Ai has noted, in the transition from Edo spaces of play to Tokyo spaces of enlightenment, there were other ruptures in the social fabric that were sutured, or remade, by new media. His assessment of this loss of community is starkly dystopian: "Compared to the neighborly world of Edo, it was a world of strangers, people who were first cut off and then reunited into an urban space by the networks of information symbolized by newspapers and the telegraph."[47]

Lest we overlook the language of the text itself, "The Thieves" was written and performed in the Fukagawa dialect, the vernacular of Edo-Tokyo urban merchant culture that would come to be adopted as the basis for standard dialect and national language reforms in the ensuing decades. The rough-and-ready speech patterns captured by Mokuami are contemporaneous with the *rakugo* performances by Sanyūtei Enchō that were transcribed in shorthand and upheld as the model for the language of unified-style prose fiction. The outsider status of "The Thieves" and its otherwise extinguished "living history" is therefore not an outlier to the origins of modern Japanese language and literature, but internal to its discursive formation.

47. Maeda, *Text and the City*, 84.

PART II

Scripting National Language

CHAPTER 4

Japanese in Plain English

From the 1870s to 1900s, debates over the "Question of National Language and Script" (*kokugo kokuji mondai*) sought to achieve a direct correspondence between written and spoken Japanese. Although these developments are often teleologically read back from the postwar ascendance of the unified style, I contend that "unification of speech and writing" did not initially emerge from a preexisting regional dialect such as the Fukagawa dialect of Edo/Tokyo per se, but from experimentation with conventional and newly invented phonetic scripts. Arguably to an even greater degree than their Western predecessors or counterparts, Japanese elites were deeply invested in the burgeoning science of phonetics and its potential to remediate the heterogeneity of dialects, spelling rules, modes of inscription, education systems, and so on. If the fragmented linguistic and social conditions of the nation-state and its colonies were to be organized under a single regime of a standardized national language, phonetic scripts would be their programming codes.

In a profound departure from the premodern cosmologies in Tokugawa thought, which held that only by stripping away the layers of latter-day corruption from written texts could the original speech and pure spirit of antiquity be revealed, Meiji reformers variously conceived of the materiality of language as that which secured meaning while remaining exterior to it. For neo-Confucian thinkers such as Ogyū Sorai (1666–1728), the proper pronunciation as well as written form of Chinese characters was necessary to ensure the rectification of names and continuity with the authority of the ancient sage-kings. For Motoori Norinaga (1730–1801), on the other hand, this phonocentrism was conversely manifested by stripping away the corrupting influence of Chinese characters and thinking (*karagokoro*) to reveal the primordial, indigenous speech and spirit (*kotodama*) of the Yamato people.[1] Thinkers from opposite ends of the Tokugawa ideological spectrum recognized the incommensurable conditions of verbal and inscriptive heterogeneity and envisioned the recovery of pure ancient speech. Yet they could not surmount the formidable obstacles to the restoration of the ideal. Consequently, theirs was a possibility foreclosed at the very moment of its conception, or what Naoki Sakai has described as the "stillbirth of the Japanese language and ethnos."[2]

1.   The critique of *karagokoro* preceded Motoori with Kamo no Mabuchi's reinterpretation of Japanese writing according to the plenitude of the supposedly indigenous "fifty sound" kana chart (*gojyū onzu*). Kamo no Mabuchi denounced the supposed superiority of Chinese origins of writing not only in relation to Japan, but also to the phonetic scripts of India and Holland:

> People also tell me, "We had no writing in this country and therefore had to use Chinese characters. From this fact you can know everything about the relative importance of our countries." I answer. "I need not recite how troublesome, evil, turbulent a country China is. To mention just one instance—there is the matter of their picture-writing. There are about 38,000 characters in common use, as someone has determined. . . . Every place and plant name has a separate character for it which has no other use but to designate this particular place or plant. Can any man, even one who devotes himself to the task earnestly, learn all these many characters? Sometimes people miswrite characters, sometimes the characters themselves change from one generation to the next. What a nuisance, a waste of effort, and a bother! In India, on the other hand, fifty letters suffice for the writing of the more than 5,000 volumes of the Buddhist scriptures. A knowledge of a mere fifty letters permits one to know and transmit innumerable words of past and present alike. This is not simply a matter of writing—the fifty sounds are the sounds of Heaven and earth, and words conceived from them are naturally different from Chinese characters. Whatever kind of writing we may originally have had, ever since Chinese writing was introduced we have mistakenly become enmeshed in it. . . . In Holland, I understand, they use twenty-five letters. In this country there should be fifty. The appearance of letters used in all countries is in general the same, except for China, where they invented their bothersome system . . . The opinion that the characters are precious is not worth discussing further. (de Bary, Gluck and Tiedemann, eds., *Sources of Japanese Tradition*, 2:13–14; *Sekai daishisō zenshū*, 54: 2–10)

2.   Sakai, *Shisan sareru Nihongo Nihonjin*, 166–210.

This impasse was renegotiated in the first decades of Meiji from many vantage points, including the new science of phonetics. National language and script could now be conceived as sharing a common basis in letters that need not have originated in "China" or "Japan" in the first place. This is not to say that modern Japanese was engaged in a belated effort to imitate or catch up with fully formed Western models, as proponents of modernization theory commonly allege. Instead, I seek to demonstrate how Western and Japanese language reformers were respectively, and in some cases concurrently, engaged in campaigns for orthography, experimental scripts, and planned or artificial languages.

Major discoveries that help decipher ancient Near Eastern writing systems also exerted a powerful influence on thinking about the origins of phonetic and ideographic scripts. Yet by the logic of nineteenth-century Orientalism, Egyptian and Chinese characters were mutually contained within the register of "hieroglyphics," and accordingly regarded as the non-Western, premodern, and figural Other to a modern, standardized alphabet. This may have set the agenda for critiques of Japanese, but it did not prevent reformers from using the term against English and other European languages that fell short of their ideals.

The earliest and most prominent example of these developments was the unofficial proposal made by Mori Arinori (1847–1889) in 1872 for Japan to adopt a simplified written form of English as the basis for national language and script. Although this is a point often missed by critics who ridiculed its impracticality, Mori's simplified English was not only predicated on the need to overcome what he called the "deranged state" of contemporary Japanese, it also leveled a broadside against the "hieroglyphie" of contemporaneous English spelling. His proposal was in fact influenced by debates in the United States about the creation of American English, which went back a century earlier to the drafting of the Constitution. As such it sheds light on a much-overlooked aspect of Japanese-Anglophone relations—namely, the extent to which Japanese activists picked up ideas in the English-speaking world. It is of particular relevance when we consider how shorthand notation and other innovations also filtered into Japan through English sources. Let us also note that subsequent language reforms in 1880s and 1890s Japan proved remarkably consistent with the underlying orthographic tenor of Mori's proposal. They included policies to limit and regularize kana, including the elimination of variant forms as *hentaigana* (irregular characters);[3] to limit the number and form of Chinese characters taught in state-run normal schools; and to create guidelines for romanization as the nation's quasi-official, "third" national script. Further waves of language reform, including a temporary revival of a movement to eliminate Chinese characters in favor of Romanization, would come in the postwar era under the very different conditions of the U.S. Occupation, when the victors drafted a new Japanese constitution.

The latter half of the chapter turns to an argument by leading Meiji philosopher Nishi Amane for the adoption of Roman letters as a fundamental precondition for "enlightened civilization" to take hold in Japan. To be more precise, he advocated orthographic reform of the alphabet to achieve a transparent relation between speech and writing. Nishi's case for Romanization appeared in his article "Yōji wo motte kokugo wo sho suru no ron" 洋字を以て国語を書する論 (Writing Japanese with the Western alphabet) in the inaugural issue of the *Meiroku zasshi* 明六雑誌 (Meiroku journal, 1874).[4] I also seek to correlate this, however briefly and impressionistically, with Nishi's *Hyakugaku renkan* 百学連環 (Encyclopedia, 1870). In this work introducing the Japanese reading public to the organized fields of knowledge in the West, Nishi gives particular prominence from the outset to the technologies of the letter. Following the general trend of Western philology, moreover, he contrasts the prevalence of universal alphabetization in the modern West against the Chinese character and other forms of ancient, and essentially fossilized, hieroglyphics. The implications for Japan's path forward could not be clearer.

3.   It is tempting to translate *hentaigana* as "perverted characters" in keeping with the punitive legislation of sexuality that simultaneously pathologized the modern configurations of homosexuality as *hentai* (perverse, deviant) during this period. On the relations between normative linguistic and sexual discursive formation in Meiji, I refer the reader to Keith Vincent's "Writing Sexuality: Heteronormativity, Homophobia, and the Homosocial Subject in Modern Japan."

4.   Befitting its critical context in canonical studies of *genbun itchi*, Nishi's article is reproduced as the second document after Maejima Hisoka's "Kanji go-haishi no gi" in Yamamoto Masahide's *Kindai buntai keisei shiryō shūsei*, 136–45. The idea for the journal was first proposed by Mori in 1873, and it was cofounded by Mori, Nishi, Fukuzawa Yukichi, Kato Hiroyuki, Tsuda Mamichi, Mitsukuri Rinshō, Mitsukuri Shūhei, Nakamura Masanao, and Suji Koji.

Mori Arinori and the Anglophone Roots of Modern Japanese

Widely reprinted as the preface to *Webster's American Spelling Book*—also known as Webster's "Blue-Backed Speller," the nineteenth century's ubiquitous textbook for teaching American children how to read, spell, and pronounce their words—Henry Steel Commager's essay "Schoolmaster to America" (1936) surveys the discovery and spread of national languages in the nineteenth century and describes conditions of linguistic heterogeneity in European history that are astonishingly resonant with Tokugawa-era Japan.

> The peoples of Old World nations had, indeed, their own language, but their language was neither common nor uniform. Everywhere on the Continent the upper-classes spoke French, and disdained the vernacular. Each region had its own dialect and each class its own idiom. So pronounced were the local dialects that Frenchmen from Brittany and Languedoc could not understand each other, or Germans from the Rhineland and Saxony, or Danes from Copenhagen and Jutland, or, for that matter, Englishmen from East Anglia and Devonshire.[5]

Commager maintains that such linguistic differences did not matter in prenational contexts, or at least mattered significantly less, due to the lack of political representation and consideration of other unifying factors such as a common religion and monarchical rule. Yet in the construction of the modern nation-state and its imagined community, establishing the uniformity of language became a pressing goal. Commager's insights into the recent origins of the nation-state presciently anticipate the central tenets of Benedict Anderson's thesis almost a half-century later. They are also transferable to the Japanese context. We need only substitute the feudal domains of Owari, Bungo, and Kii for the provinces of France, Germany, Scandinavia, or Britain for this statement to apply with equal force to pre-Meiji Japan.

---

5. Commager, "Schoolmaster to America," 7.

Commager's categorization of the origins of the nation-state further speak to the recuperative projects of German Romanticists, Japanese Nativists, and others to uncover the true nature of the folk in a continuous national language and literature. Commager continues:

> A nation needs not only a common language; it needs, even more, a common past, and a sense of that past. Every European state-maker of modern times has recognized this. Thus, Bishop Gundtvig in Denmark devoted his volcanic energies to the editing of ancient Danish ballads, the writing of national histories and national songs, the resurrection of the national past. Thus, in Germany Schlegel and Stein and Savigny and the brothers Grimm recreated the German past in order to create a German future. . . . Thus, Ernst Renan emphasized for France "the common memories, glories, afflictions, and regrets"; thus in England John Stuart Mill concluded that the most important ingredient in nationalism was "the possession of a national history and community of recollections." And our contemporary, Sir Ernest Baker has put the matter succinctly: "A nation is not the physical fact of one blood but the mental fact of one tradition."[6]

For nations new and old, this mental fact began with the invention of national language and literature, which in turn depended on scripts. By the same logic as Commager's essay, a material history of writing practices, educational regimes, and so on must be grounded in these essential building blocks of language.

Before becoming Minister of Education (1885–1889), Mori Arinori was the first ambassador to the United States, serving from 1871 to 1873. It was in this capacity that Mori issued his proposal for adopting a simplified form of English as the basis for Japan's national language. The proposal was written as an open letter to William Dwight Whitney, a specialist in Sanskrit and comparative philology at Yale University, and subsequently published in the American newspaper the *Tribune* and in the Yokohama-based Japanese newspaper, the *Japan Weekly Mail*.[7]

---

6. Ibid., 9.

7. See Mutō Teruaki's useful overview in "Mori Arinori's 'Simplified English.' " Mutō cites articles by three American respondents (Coates Kinney, M. G. Upton, and E. E. Hale) to Mori's proposal and provides a tentative analysis of the "diglossic" linguistic condition in Meiji. Ivan Hall's *Mori Arinori* duly notes the popular ridicule Mori's proposal garnered in Japan from the foreign community's English press.

Some twenty-three leading educators, scholars, and intellectuals in the United States responded to Mori, which he edited and published as *Education in Japan* (1873).

Although Mori is sometimes grouped together with Maejima Hisoka as a would-be abolitionist of Chinese characters, this is misleading on three counts. First, both men objected not only to *kanji*, but to the entire cultural complex related to Chinese learning. In this sense, labeling them proponents of "the elimination of Chinese characters" (*kanji haishiron*) does not go far enough. They regarded China as a discredited model and therefore no longer suited for Japan to emulate. Second, unlike Maejima, Mori's objective was to create a unified national language and script by replacing the heterogeneity of spoken Japanese dialects with a standard dialect commensurate with a perfected English orthography. Third, although Maejima's proposal was nominally addressed to the shogunate, which ignored it, and only publicly reprinted after the shogunate's demise, Mori's initial audience was neither the Meiji government nor the Japanese people, but leading American scholars and thinkers capable of throwing their authority behind him. It was the first of many surprising points of connection between language reform in Japan and the Anglophone world that must be carefully reexamined.

Mori's proposal to abandon the heterogeneity of Japanese and adopt simplified English was integral to his views on achieving national destiny through the social Darwinian notions of fitness espoused by Herbert Spencer.[8] Mori's views on national-imperial education were deeply influenced not only by Spencer, but also by Thomas and Matthew Arnold in England and the Schrebers in Germany. In keeping with the philosophy of national education and his understanding of bodily and moral discipline, Mori sent his young protégés such as Isawa Shūji to newly founded Pestallozian institutions such as the Bridgewater Normal School

---

8. See Anderson's "The Foreign Relations of the Family State," 21–69. Anderson explores the epistolary correspondences between Mori and Spencer in matters of physical and moral education and their exchanges on "the inscription of capital and the social and physiological body" (25). I should note that Anderson does not address the question of national language or Mori's letter to Whitney, which preceded his other writings.

in Massachusetts and Oswego, New York.[9] It is not altogether surprising in light of his early forays into language policy that as the first minister of education, Mori presided over the introduction of normal schools, military-inspired school uniforms, and quasi-military indoctrination in moral and physical education.[10] Mori's attempts to implement Spencerian ideas in Japan were consistent with developments in Anglo-American education such as the notion of "muscular Christianity" associated with Theodore Roosevelt, and the moral and physical, but decidedly anti-intellectual, public school education in Britain espoused by Thomas Arnold.

Another major figure who shaped Mori's views on national language reform was Noah Webster. In his *Dissertations on the English Language* (1789), Webster called for a linguistic unification of the new nation akin to the ratification of the United States Constitution which took place in March of the same year: "Let us then seize the present moment and establish a *national language*, as well as national government."[11] Webster insisted on establishing a uniform national tongue based on an American rather than a British vernacular, which would be vouchsafed by a simplified spelling system. Unnecessary silent letters and multiple-letter combinations representing the same phonetic values would be regularized: "laugh" would become "laf," "grief" would become "greef," "machine" would become "masheen," and so on. Ironically for the contemporary reader, Webster's *Dissertations* was published in colonial-era typography including now obsolete features such as the long *s* and the *ct* ligature. The 1817 reprint of Webster's *American Spelling Book*, on the other hand, betrays no variation from contemporary usage. The disappearance of these letters clearly took place in the intervening years in the early nineteenth century. In a manner of speaking, English had its own "hentaigana" that preceded orthographic standardization in the nineteenth century.[12]

---

9.   As Ivan Hall notes, Mori had spent the year 1867 in upstate New York with the Brotherhood of New Life, a religious colony led by ex-Swedenborgian spiritualist and sexual mystic Thomas Lake Harris. See *Mori Arinori*, 106–28.

10.   See Mori's *Rinrisho* in Ōkubo, ed., *Mori Arinori Zenshū*, 1:419–54, which he intended "to set a standard" for moral judgments.

11.   Webster, *Dissertations on the English Language*, 406 (emphasis in original).

12.   Kana were written with far greater variation than the calligraphic hands or typographic fonts of these now-obsolete English letters. Nevertheless, I want to underscore

It is curious that to date there have been no scholarly investigations of the affinities between Mori and Webster, or related works such as Benjamin Franklin's 1768 proposal for a reformed English alphabet and spelling rules.[13] More baffling still is Whitney's silence on the matter in his reply to Mori, considering that he had been the editor of *Webster's American Dictionary* (1864) and the founder and first president of the American Philological Association (1869). Mori, however, touches upon the historical basis for his argument when he points to the unfulfilled promise of Anglophone reform: "I propose merely to complete what all English and American Lexicographers, from Dr. Samuel Johnson, down to the authors of the changes contained in the latest editions of Walker's, Webster's, and Worcester's dictionaries all commenced but timidly abandoned."[14] It was this inconsistency in spelling that led Mori in his letter to Whitney to lament not only the hieroglyphic quality of written Japanese, but of English orthography. He boasts that his proposal for standardized English would "mak[e] the language actually what it claims to be—phonetic—instead of hieroglyphic on a phonetic basis, which is what it now really is."[15]

Mori further reiterates Noah Webster's bold visions for this new American tongue when he claims it will be a national project far removed from the shores of the Anglophone world. Mori does not presume to change American or British English, but rather to carry out "an adaptation of the English language to the necessities of a foreign nation of forty million souls, separated by thousands of miles from the English-speaking nations, and which affords an entirely free field, for the introduction of a new language; there being no obstacle whatsoever within the Empire itself."[16] For his part, Webster saw the American continent and the immigrant peoples populating it as a unique opportunity for breaking away from "blind imitation" of England:

---

that the uniform spelling and appearance of Roman letters in English is a product of the same nineteenth-century reforms that led to the standardizations of Japanese scripts.

13.   Franklin's revised alphabet, which had a direct influence on Webster's, involved the elimination of six letters of the alphabet he deemed redundant (*c, j, q, w, x,* and *y*) and introduction of six new ones.

14.   Katō and Masao, *Honyaku no shisō*, 320.

15.   Ibid., 320.

16.   Ibid., 320.

[W]ithin a century and a half, North America will be peopled with a hundred millions of men, all speaking the same language. . . . Compare this prospect, which is not visionary, with the state of the English language in Europe, almost confined to an Island and a few millions of people; then let reason and reputation decide, how far America should be dependent on a transatlantic nation, for her standard and improvements in language. . . . These causes will produce, in a course of time, a language in North America, as different from the future language of England, as the modern Dutch, Danish and Swedish are from the German, or from one another: Like remote branches of a tree springing from the same stock; or rays of light, shot from the same center, and diverging from each other, in proportion to their distance from the point of separation.[17]

Although one must always be wary about the specter of comparison in East-West relations, one can say that both Webster and Mori saw geographical remove, coupled with a revolutionary self-determination, as imparting a special destiny to their young nations.

Needless to say, Mori's sanguine proposal gives no hint of how to *practically* implement this script reform as a spoken language, much less how to combat the existing heterogeneity of Japanese dialects and writing styles even within the historical confines of the Japanese archipelago. The main thrust of Whitney's reply seizes upon this dilemma to point out that one could not possibly impose such sweeping changes in daily life and language without incurring massive popular uprisings or inviting a potentially massive economic and cultural catastrophe. He recommends accepting English as is, and then gradually teaching it as a second language, a view that Ivan Hall speculates may have been shaped by Whitney's own knowledge of the tensions in British-occupied India.[18] Japan's forced linguistic and cultural assimilation of Koreans from 1938

---

17. Webster, *Dissertations on the English Language*, 21–23.

18. Katō Shūichi's "Meiji-shoki no honyaku" (Translation in the early Meiji era) notes critiques by Whitney and Baba Tatsui about the imposition of a foreign language upon the native population (*Honyaku no shishō*, 349). Ivan Hall likewise cites Baba's condemnation of Mori's proposal (*Mori Arinori*, 194). Parenthetically we might note that there is a curious resonance of Whitney's recommendation to accept English *as it is* with the compositional ethos of shorthand realism to "transcribe things just as they are" (*ari no mama ni utsushitoru*).

(*naisen ittai*) would generate similar conflict and a legacy of deep resentment.

Did Mori truly believe that simplified English would ipso facto solve the problems of fractured spoken tongues in the Japanese archipelago? I need not try to defend or minimize the tremendous leap of faith this requires, but what is often missed by those heaping abuse on Mori's head is that his desire to implement a purely phonetic English orthography placed him squarely amidst the international language theorists and imperial language planners of the era. Aspiring to become linguistic equivalents to world standard time and the metric system were numerous artificial or "planned" world languages: Johann Schleyer's Volapük (1880), L. L. Zamenhof's Esperanto (1887), Giuseppe Peano's Latino sine flexione (1903), and the reformed Esperanto known as Idiom Neutral or "Ido" (1907). These languages did not emerge spontaneously from a linguistic community, but through the methodical planning of linguists. Ironically, though principally intended as international spoken languages, their perfectly regularized grammars and vocabularies were almost entirely worked out, like Mori's, on paper. Perhaps best capturing their utopian dreams are Volapük and Esperanto, whose names respectively mean "world speech" (*vola*, from the French and *pük*, a modification of the English word "speak") and "to hope or aspire."

Despite the disappointing failure of Mori's simplified English, more direct echoes of simplified English would persist for decades to come, including Charles Kay Ogden's *Basic English: A General Introduction with Rules and Grammar* (1930), which was promoted in China and elsewhere by I. A. Richards.[19] Still, we need not look so far ahead to encounter similar expressions of dissatisfaction with English. In 1909, American spelling reform advocate Thomas Lounsbury wrote a statement that sounds strikingly similar to Mori's: "Every member of the English race has to learn two languages . . . the one he reads and writes; the other he speaks."[20] Lounsbury likewise remonstrated with what he saw as the ill-founded claims of British historicists to retain existing spellings as ties to a more

---

19. Basic English, with its minimalist range of expression and imperialist ambitions to become the postwar lingua franca, was the inspiration for "Newspeak" in George Orwell's *1984*.

20. Lounsbury, *English Spelling and Spelling Reform*, 279.

ancient and pure past. He indicates a number of false etymologies that arose from misspellings by printers, copyists, and passing fashions. Spelling rules, he avers, were historically ad hoc and more often than not established by individual printers rather than scholars or governments during those historical periods before standardization. Consequently, the haphazard and multiple spellings that persisted into the nineteenth century did not provide consistent rules for observing a single historical precedent. Lounsbury delights in pointing out the majority of Shakespeare's extant spellings are, in fact, closer to American than British custom. A comparable situation existed in Japan with the proliferation of kana and even alternative Chinese characters until the standardization campaigns of Meiji got under way. Thus, he concludes, spelling rules in their present form are maintained not by logic or reason, but by sentiment. Much as Confucian schools of thought in East Asia held that the correct form of a character or word was intrinsically tied to its true meaning—i.e., the rectification of names—so, too had English linguists and philologists since Samuel Johnson sought to preserve those affective traces of the past encoded as spelling rules. The universality of these issues underscores the extent to which the phonetic turn from the mid-nineteenth century made it possible to reconceive of language beyond its familiar roots and branches.

The frequent reference to hieroglyphics in nineteenth century thought also requires some contextualization. The discovery of the Rosetta stone in 1799 and its deciphering by French scholar Jean François Champollion in 1822 enabled Europeans to finally grasp the cofiguration of sounds and meanings in ancient Egyptian. Of course, this did next to nothing to impede popular beliefs regarding the sacred mysteries of hieroglyphic scriptures,[21] nor did it prevent the term "hieroglyphic" from having an extremely long shelf life as the diametrical opposite to alphabetic or phonetic writing.[22] Less than a decade after Champollion's deciphering of the Rosetta stone, Hegel's *The Philosophy of Mind* (1830) posited the

---

21.   *The Book of Mormon* was published in 1830 from "Egyptian-Hebraic" inscribed golden plates found in 1823 under a glacier in upstate New York by the founder and prophet of the Church, Joseph Smith. In language befitting the times, Smith called the script "reformed Egyptian" (*Book of Mormon* 9:32).

22.   For instance, in his discussion of montage as an underlying essence in traditional Japanese culture in "The Cinematographic Principle and the Ideogram" (1929), Sergei Eisenstein uses the term interchangeably with "ideograph."

superiority of the alphabet as the sign system that maintains the closest relation to speech:

> While on the subject of spoken language (which is the original) language, we can also mention, but here only in passing, *written language*; this is merely a further development within the *particular* province of language which enlists the help of an externally practical activity. Written language proceeds to the field of immediate spatial intuition, which it takes and produces signs. More precisely, *hieroglyphic script* designates *representations* with spatial figures, whereas *alphabetic script* designates *sounds* which are themselves already signs. Alphabetical writing thus consists of signs of signs, and in such a way that it analyzes the concrete signs of spoken language, words, into their simple elements and designates these elements.[23]

Of course Hegel did not confine usage of the term "hieroglyphics" to Egypt. The other country he targets for hieroglyphic writing that bleeds over into opacity of speech is China. It is the very lack of alphabetic writing, he opines, that makes the tonal system in Chinese necessary:

> The imperfection of the Chinese spoken language is well-known; a mass of its words have several utterly different meanings, as many as ten, or even twenty, so that, in speaking, the distinction is made noticeable merely by stress and intensity, by speaking more softly or crying out. Europeans beginning to speak Chinese stumble into the most ridiculous misunderstandings before they have mastered these absurd refinements of accentuation. Perfection here consists in the opposite of that *parler sans accent* which in Europe is justly required for cultivated speech. Owing to hieroglyphic written language the Chinese spoken language lacks the objective determinacy that is gained in articulation from alphabetic writing.[24]

Although Hegel recognizes the lack of a purely phonetic system in the West, he holds alphabetic writing with its "objective precision" superior to the confusion of Chinese. It goes without saying these blanket assumptions about the inherent advantages of alphabetic writing are continually

---

23. Hegel, *The Philosophy of Mind*, 195–96.
24. Ibid., 196–97.

beset by the problems of homophones, arbitrary spelling rules, and so forth. In Japan, as well as Europe and America in the latter half of the nineteenth century, phonocentrism would be sorely challenged by the on-slaught of mechanical and standardized models developed by propo-nents of language and scripts reforms.

Nishi Amane's Case for Romanization

If we take the *Meiroku Journal* on faith, then in the beginning was not the word, but the letter. Published in the inaugural issue of the *Meiroku Journal* in March 1874, Nishi Amane's article "Writing Japanese with the Western Alphabet" was sympathetic with Mori's diagnosis of the need for a unified national language, but unwilling to endorse simpli-fied English as the solution. While praising the Imperial Restoration's reform of government and law, Nishi decried the prevalence of ignorance that the Meiroku Society was formed to combat. Basing his own pro-posal on the materiality of the letter, Nishi presented an only slightly less shocking alternative to Mori when he argued for the wholesale adoption of Romanization. Indeed, he argues that it is the materiality of letters themselves that permits mental or enlightened civilization to come into being.

> If we now just consider the import of the words "science, the arts and let-ters" in the motto of our society, what we call science and the arts emerge after letters [*bunshō*] exist. Without letters, how can there be science and the arts? The ancients also said that literature is the means for understand-ing the Way. In our letters at present, however, it is improper for us to write as we speak as well as improper to speak as we write since the grammars of speech and writing in our language are different.[25]

25. Braisted, ed., "Writing Japanese with the Western Alphabet," in *Meiroku Zasshi*, 5. Although the Japanese word *bunshō* literally translates as "writing" in the basic sense of composition or sentence structure, it is evident from the context of the paragraph that it is closer in meaning to letters in the dual sense of humanistic literature, as per the motto of the Meiroku Society, and the building blocks of written language, as per his comments about the incommensurability of speech and writing.

In the first part of his argument, Nishi divests "letters" of their storied precedents in classical or Nativist learning. He abandons the ritual potency of "Chinese letters" (*kanji*) and native particularism of "Japanese letters" (*waji*) in favor of romanization, which he calls "European letters" (*yōji*). Only by adopting Western letters, runs the logic of Nishi's argument, could Japan effectively assimilate Western learning.

Despite somewhat contradictory initial statements on whether to endorse the creation of a vernacular written style or adhere to patterns of classical grammar, Nishi maintains that "since writing and speech will follow the same rules, what is appropriate in writing will also be appropriate in speaking. That is, lectures, toasts, speeches before assemblies, and sermons by preachers may all then be recited as they are written and recorded as they are read" (*iu tokoro kaku tokoro to sono hō wo dōsu, motte kaku beshi motte iu beshi, sunawachi, rekichua tōsuto yori kaigi no supīchi hōshi no setsuhō mina sho shite tōsu beku yonde sho subeshi*).[26] Here is the idea of one-to-one commensurability with an added emphasis on the values of public speaking being propounded almost a decade prior to Takusari Kōki's adaptation of phonetic shorthand.

Nishi articulates a compromise between ornate and vulgar styles, or what might be called the "mixed style" (*gazoku setchū*). Providing a chart of adjectives, nouns, and verbs in Edo and Kyoto speech patterns, Nishi uses Romanized words with a dot (.) to show letters omitted in vernacular speech and a circumflex (ˆ) for morphological changes from classical grammar to the mixed style. Although by his own admission this style was far from complete, he pushed the limits of existing graphic regimes to represent its verbal features.[27] Among the many advantages he anticipated were the transformations of Japanese manuscript and woodblock print culture into a new typographic regime, and the creation of an educational system equal to those in the West. Nishi likewise made proposals to forestall the opposition of traditional merchants whose livelihood

26. Braisted, ed., *Meiroku Zasshi*, 5.

27. The style of Romanization was his own, not Hepburn's or Rosny's, which he labels imperfect approximations of vulgar speech. His system appears similar to Tanakadate's much later invention of the Nippon-shiki in its economy of letters (*si* rather than *shi*) to avoid noise or interference in the channel of transmission. Perhaps coincidentally, the use of dots and circumflexes was also a feature of some nineteenth-century shorthand notation systems.

selling brushes, ink, and traditional paper would be adversely effected, as well as Nativist and Confucian scholars whose prestige, not to mention livelihoods as teachers, were threatened.

Nishi recognized a lack of consensus even amongst the elite intellectuals in the *Meiroku Journal* as evidence of a deeper intransigence that would require further debate to overcome:

> Recently, in publications such as this journal, it has been fairly standard to write with a mixture of [Chinese ideographs] and kana syllabary. Even so, these publications have achieved no stylistic unity, as from time to time they mix Chinese and Japanese grammatical forms. Those opposed to the Japanese classical scholars, therefore, desire ultimately to write directly in today's vulgar tongue [*zokugo*] and to abandon completely the system of postpositions.[28]

Nishi identifies the lack of standardization and the tension between classical written and vernacular spoken forms as dominant concerns amongst his peers. He may well have had in mind such works as Fukuzawa Yukichi's *Gakumon no susume* (Encouragement to learning, 1872–1876), which was then midway through its serialization and which Nishi discussed in the second issue of the *Meiroku Journal*. Fukuzawa, who did not contribute to the *Meiroku Journal* until the twenty-first issue, utilized a plain style accessible to a broad-based readership less schooled in Chinese characters or academic styles of writing. To Nishi's credit, the inaugural issue of the *Meiroku Journal* sold about three thousand copies, ensuring him a respectable audience amongst Meiji elites, including the linguist credited with the term "unified style" itself. As Nanette Twine points out, "Mozume Takami (1847–1928), who became an influential advocate of colloquial style with the publication of his essay *Genbun itchi* in 1886, recalled that it was the influence of Nishi's essay and his subsequent visit to discuss the matter further which first sparked his interest in the subject."[29]

---

28. Braisted, ed., *Meiroku Zasshi*, 10.

29. Twine, *Language and the Modern State*, 119. Twine also points to Mozume's comparison of *genbun itchi* with the benefits of the modern postal system (216). In fact, Mozume makes a historical comparison between Meiji postal discourse in reshaping national communications yet recognizes that the heterogeneity of spoken dialects and written forms of language have not changed that much since the days of Jippensha Ikku's

A final aspect of his argument anticipates the discourse of realism and its relation to phonetic scripts in its call for modes of scientific observation to be adopted in Japanese language as well as other technological endeavors:

> I have pondered the fact that the Europeans now lead the world. If one considers this from the point of view of reason, their race has achieved greatness by piling the lesser results of their minute observations. They even view the vastness of the universe in terms of the falling apple. They guide legions by beginning with the training of but one soldier. They depend upon nothing more than the expanding power of steam to send ships across the four seas. Their transmission of electricity to the four continents is derived from just observing a humble kite. And similarly nothing has contributed more to their world preeminence in science, the arts, and letters, than the twenty-six letters of the "ABCs."[30]

Whether we agree with Nishi's rosily Eurocentric assessment, it is important to clarify that he was not endorsing the alphabet as a spiritual essence of the West—that is, as the basis for Western phonocentrism—but as a practical technology in advance of the imperative of shorthand to "write things down just as they are" (*ari no mama ni utsushitoru*).

By way of conclusion, let me turn to another work in which Nishi had already explicated his views on materialism, *Hyakugaku renkan* (Encyclopedia). With its title literally meaning "the links of the hundred sciences," Nishi's text proceeds somewhat fitfully between sets of binary relations that establish the order of Western branches of knowledge. Further, as Kōsaka Masaaki observed, "Nishi grouped learning horizontally into science and technology, vertically into general (universal) and particular (specific). By dividing the particular in turn into intellectual and physical he completed the structure of his 'Encyclopaedia.'"[31] Yet the

---

Tōkaidō Hizakurige (see Mozume's "Genbun itchi" in *Mozume Takami Zenshū*, 3:8). See also Yamamoto's *Genbun itchi no rekishi ronkō*, 94–95. Based upon Ueda Kazutoshi's *Kokugo no tame* (in Hisamatsu, ed., *Meiji bungaku zenshū, vol. 44*), it is clear there remained some discrepancy over nomenclature, as Ueda refers to the new vernacular style as both *genbun itchi* and *genbun itto*.

30. Braisted, ed., *Language and the Modern State*, 15.
31. Kōsaka, *Japanese Thought in the Meiji Era*, 102.

branching of these axes scarcely do justice to the a priori importance of the letter. Nishi insisted that the *alphabetic* list of entries provided the correct order by which the exploration of "science and art" (*gakujutsu gigei*) should be conducted.[32]

Nishi's *Encyclopedia* is perhaps better understood as an indexing of disciplinary knowledge rather than a systematic introduction to them. In making this claim, I simply want to emphasize that Nishi foregrounded the materiality of writing in the production of knowledge even when it came at the expense of a more orderly or normative philosophical sequence. In rapid succession, the *Encyclopedia* marks the branching of arts and sciences, theory and practice, then mechanical and liberal arts. It is at this third branch that his argument enters into the material and institutional underpinnings of mental civilization, from typographic print culture, including the newspaper, to institutions of higher learning. The divisions unfold in such a way that variations in the Indo-European languages, such as "father, vader, père, pater, pitar," appear side by side with the technical language of "end, means, measure, and medium" and "mechanical instrument." It is a remarkable statement on the face of it, placing media on the same level as patriarchy. By the same token, "literature" (*bunshō*) begets knowledge for Nishi as much as institutions such as "school, university, academy."

In the section on language, the opposition of "hieroglyphic" (*keishō moji*) and "letter" (*onji*) are noted in passing, but disappointingly not investigated at length. In Nishi's original manuscript, however, this binary is graphically sketched out with figural and phonetic markers comparing the different scripts of the Egyptian and Chinese, as well as the ancient "Mexican" (Aztec) and Phoenician. These rough notes that bring together the various scripts of world civilizations in a common register reflect the degree to which Nishi was keenly aware of Japan's lack of universal alphabetization. His advocacy for Roman letters as the basis for Japan's national script was hardly predicated upon ignorance of the diversity of human writing, or a knee-jerk acceptance of all things Western. What

---

32. It was only after the reform of Japanese education by the Ministry of Education under Mori that the *iroha* organization of the kana syllabary was steadily replaced by the more alphabetical *a-i-u-e-o* order that is universally in use today. See Mabuchi Kazuo's *Gojyūonzu no hanashi*, 14–28.

Nishi could not anticipate, however, was the rise of experimental phonetic scripts, which made his and Mori's critique of alphabetic writing seem tame by comparison. Nor could he have divined the reorganization of human hands, ears, eyes, and mouths into so many moving parts of writing technology.

CHAPTER 5

Phonetic Shorthand

The brain with its supplementary organs resembles a camera of the latest and most approved type, namely the stereoscopic instrument with two sets of lenses corresponding to the eyes, which gives to the object photographed its natural form. . . . The brain not only resembles an apparatus for taking pictures, but it also appears to be very much like that still more modern invention, the phonograph. . . . All these photographs and phonographic records are stored away, and if perfectly taken, they can be recalled, at any time, before the mind's eye or the mind's ear, and constitute what we call memory.

—E. A. Sturge, address to the Imperial Educational Society, 1904

In an address on "Language Study" to the Imperial Educational Society reprinted in the July 1904 issue of *Taiyō* 太陽 (The sun), Ernest Adolphus Sturge (1856–1934), an American physician and proponent of Romanization, drew upon the latest scientific findings to argue for the superiority of alphabet learning.[1] Sturge informed his esteemed audience that the mind stores images and sounds much as the imprinting of light and sound waves are recorded on photographic plates and phonographic records. These, he averred, may fade or become damaged over time in similar respects: "By means of the auditory organs the waves of sound write their impression upon another brain film, which with the aid of the vocal organs can be more or less perfectly reproduced as by the phonograph."[2] In short, Sturge staked his claims upon the technical capabilities of late nineteenth-century media. Of course, he was hardly alone in conceptualizing the human mind as a psychic surface of inscription, or *encryption*, hooked up to ears, eyes, and mouths to record and retransmit life's

1. Sturge, "Language Study," 6–7.
2. Ibid., 7.

datum.[3] Regardless of its actual scientific merits regarding the hardwiring of the human brain, his lecture underscores a familiar slippage across technological registers that had emerged two decades earlier in Meiji culture in the rhetoric of shorthand as phonography and verbal photography. In fact, shorthand was pivotal in transporting enunciations from speech acts to the authenticity of typography while embracing a rhetoric of high fidelity commonly thought of today as belonging exclusively to mechanical (and now digital) devices.

It is unclear to what extent Sturge was aware of the ongoing debates surrounding Japanese language and script reform. Nevertheless, the invitation for him to speak at the Imperial Education Society and subsequent publication of his remarks confirms that there was still a receptive audience for drastic reforms in the final decade of Meiji. Knowingly or not, Sturge echoed Maejima's, Mori's, and Nishi's claims that Romanization will free Japanese from the mnemonic burden of "hieroglifphics" [*sic*], and facilitate Japan's move closer toward the Anglophone and Western world. His argument is closest to Nishi's in stating that Romanization would drastically streamline the cumbersome process of Japanese typography:

> A Japanese printing office, as compared with ours, is most inconvenient. The selecting of the necessary characters for a daily paper, and the sorting of them again after the printing has been done, is a great task. If instead of these, you should choose the Roman letters, you could find them all *close at hand* in a small convenient case.[4]

This turn of phrase so conducive to the Heideggerian notion of standing-reserve allows for the infinite possibilities of language to be contained within combinations of twenty-six letters, ten numbers, and assorted punctuation marks. It is fitting, too, that the case is no longer a figure of speech, but a literal representation of the Western typographer's tool box.

---

3.   Kittler outlines similar premises for the human brain as a site of data storage and retrieval in Freud's "Project for a Scientific Psychology" (1895) in *Gramophone, Film, Typewriter*, 37–38. Freud, as Kittler is fond of reminding us, depended on the infallibility of his own "phonographic" memory and employed the talking cure as the basis for a psychoanalytic process that gained access to the most latent and buried *communiqués* from the unconscious mind.

4.   Sturge, "Language Study," 10 (my emphasis).

As we have already observed, the project of unifying Japanese took place amidst volatile contestations of national and imperial languages throughout the world, including the rise of experimental scripts and planned languages. This chapter primarily seeks to historicize the origins and adaptation into Japan of phonetic shorthand. It is a topic that has been largely consigned to the periphery of modern Japanese literary and linguistic studies, particularly when compared to kanji, kana, and even Romanization. Mainstream proponents of phonetic shorthand in the West and Japan believed it capable of faithfully reproducing speech *as is*, not merely as its approximation. It could then be overwritten, palimpsest-like, by more conventional writing systems. They maintained that the incommensurability of speech and writing could not only be surmounted, but also transcended, by shorthand.

Phonography and Verbal Photography

The putative origins of shorthand do not begin in the nineteenth century, but with the symbolic or figural writing buried beneath the surface of ancient scripts, including Egyptian (fourth century BCE), Greek (second century BCE), and Roman (first century BCE). In most popular accounts, Western shorthand begins with Cicero's freed slave Marcellus Tullius Tiro. He is credited with a notation system called the Notae Tironianae, which consisted of several thousand symbols representing word-stem abbreviations (*notae*) and word-ending abbreviations (*titulae*). The Tironian system was utilized with small modifications in literate Europe until the Middle Ages. For the Romans, shorthand maintained a conduit between oratory and history that was essential to the conduct of state and glory of Rome. Up to forty shorthand scribes might be stationed throughout the Senate, and their accounts would be compared and combined into a single official record. Consequently, Roman shorthand was always understood to be *cursory*, rather than a precise duplication of the spoken word.

The Romans practiced shorthand by using a stylus on a tablet covered with a layer of wax. The stylus was fashioned with a sharp edge of ivory or steel and a flat metal handle to smooth the wax and erase previous inscriptions.[5] Records first etched in wax would then be transferred into proper Roman letters and onto paper to make them official. Nevertheless, there was never a presumption that shorthand was anything more than a provisional and approximate representation. By contrast, the metal stylus working across a wax medium introduced in 1877 with Edison's invention of the phonograph was a technology already firmly based upon the principles of precise recording and lossless transmission.

Shorthand practices in the eighteenth and early nineteenth centuries prior to Pitman were so widespread that it is difficult to isolate any single individual or notation system as the primary agent of its dissemination. The most salient characteristic was continued reliance on abbreviations and symbolic markers. This prephonetic history does not lack for luminaries. To cite but two prominent examples, James Madison recorded the 1786 Constitutional Congress using an unspecified shorthand method. In his youth in the 1830s, Charles Dickens studied Thomas Gurney's *Brachygraphy* (1750) and became one of the most sought-after reporters on the English political speech circuit. In an uncanny resonance with the transmission of shorthand into the equally new genre of the novel in Meiji Japan, Dickens made the shift from shorthand reporter to serial novelist by publishing prose sketches accompanied by illustrations in *Sketches by Boz* (1835–1836) and the *The Posthumous Papers of the Pickwick Club* (1836).

Pitman studied Samuel Taylor's so-called geometric shorthand in 1829,[6] before developing his own phonetic method, first published as *Stenographic Soundhand* in 1837 and renamed "phonography" in 1840. Pitman is in fact associated with three major developments in phonetic scripts: phonography, phonotypy, and the first International Phonetic Alphabet (IPA) in 1886.[7] Pitman portentously characterized his phonography as "the alphabet of nature," one whose arrangement of sounds was not based on arbitrary graphic markers, but determined by their mode and

---

5. The instruments were also used to record in blood another momentous event: Julius Caesar, who was himself adept in shorthand, was in fact, stabbed to death by senators armed not with knives or daggers, but shorthand styli.

6. Taylor's system, based on nineteen simplified letters, was invented in 1786 and was the most widespread in use in the United Kingdom prior to Pitman's.

7. I am grateful to John Whitman's "Transcription: The IPA and the Phonographic Impulse," unpublished manuscript (2004) for bringing these issues of my attention. For further reading on the history of the IPA, see Albright, "The International Phonetic Alphabet."

order of vocalization. Hence *p* and *b* stand in relation to one another, instead of becoming lost in the usual chaotic jumble of ABCs. Short vowels were indicated with a light dot or dash, while long vowels were marked with a heavier stroke (figure 5.1). Writing in straight and curved lines was intended to produce a smooth, continuous motion so as never to lose time or data by lifting the hand off the page. Moreover, the curvatures of the line were meant to correspond to the natural shape and movement of the human hand, which was indicated in shorthand manual illustrations of the craft. Phonography promoted the notion of automatic writing based on an unprecedented recoordination of ears, hands, and eyes with pen or pencil and paper. Punctuation in the Pitman system also introduced its share of novelties: there was a combination of familiar and new markers, including what would today be called "emoticons," or *emoji* (literally, in Japanese, "picture-words"). A downward spiraling curlicue was used to designate a smile, although Pitman warned it should be reserved for casual correspondence only. In the interests of hand speed and facility of use, the comma was left alone, whereas a small *x* was used in lieu of a period.

A large field of competing shorthand systems arose in Europe based on phonetic innovations. In Germany, F. X. Gabelsberger's 1834 system linked shorthand and musical notation by placing phonemes into scales. In France, Émile Duployé developed a syllabic system in 1862 similar to Pitman's with consonants figured in strokes and vowels represented by small circles. Among the most radical departures was Italian Antonio Michela-Zucco's modified piano keyboard with six white and four black keys invented in 1860 and adopted in 1880 for use in the Italian Senate.[8] Although it is beyond the scope of this book to investigate them in depth, national and regional variants of shorthand cropped up throughout the world and would persist well into the late twentieth century as a manual skill essential to commerce, law, and government.

The breakout success of Pitman's shorthand enabled him to pursue the related typographic project of phonotypy and to promote other aspects of phonetic reform through the Phonetics Institute he founded in the 1840s. By all accounts, Pitman was enough of a political pragmatist to recognize that despite trumping all comers with a practical and widely adopted, not to mention widely imitated system, he did not actively seek

8. The system was updated with transcription software and MIDI technology in 2003.

(21.) **TABLE OF CONSONANTS.**

	Letter.	Phonogram.	Examples of its power.		Name.	Phonotype.
Explodents.	P	\	ro*p*e	*p*ost	pee	p, *p*
	B	\	ro*b*e	*b*oast	bee	b, *b*
	T	|	fa*t*e	*t*ip	tee	t, *t*
	D	|	fa*d*e	*d*ip	dee	d, *d*
	CH	/	e*tch*	*ch*est	chay	ç, *ç*
	J	/	e*dg*e	*j*est	jay	j, *j*
	K	—	lee*k*	*c*ane	kay	k, *k*
	G	—	lea*g*ue	*g*ain	gay	g, *g*
Continuants.	F	\	sa*f*e	*f*at	ef	f. *f*
	V	\	sa*v*e	*v*at	vee	v, *v*
	TH	(	wrea*th*	*th*igh	ith	ɪ, *θ*
	TH	(	wrea*the*	*thy*	thee	ɖ, *ð*
	S	)	hi*ss*	*s*eal	ess	s, *s*
	Z	)	hi*s*	*z*eal	zee	z, *z*
	SH	/	vi*ci*ous	*she*	ish	ʃ, *ʃ*
	ZH	/	vi*si*on	*j*e (Fr.)	zhee	ʒ, *ʒ*
Nasals.	M	⌒	see*m*	*m*et	em	m, *m*
	N	⌣	see*n*	*n*et	en	n, *n*
	NG	⌣	lo*ng*		ing	ŋ, *ŋ*
Liquids.	L	/ up	fa*ll*	*l*ight	el	l, *l*
	R	\ / up	fo*r*	*r*ight	ar, ray	r, *r*
Coalescents.	W	⌐ up		*w*et	way	w, *w*
	Y	⌐ up		*y*et	yay	y, *y*
Aspirate.	H	/ ⌐ up		*h*igh	aitch	h, *h*

5.1 Sound chart from Isaac Pitman's *Phonography*, 1888. Photograph by the author.

REDUCED FACSIMILE OF THE TOP PORTION OF THE FIRST PAGE OF
" THE PHONETIC NEWS "

5.2 Isaac Pitman's *Fonetic Nuz*, January 6, 1849. Photograph by the author.

to overthrow the alphabetic tradition. Or at least not overnight. As J. Kelly notes, "Pitman came to see phonography as merely the first stage of a larger campaign having as its object the replacement of the traditional writing-system by a more rational system, related in its phonetic basis to the phonographic system."[9] The compromise script that emerged from these efforts was phonotypy, which combined old and new letters of the alphabet with the intention of maintaining a one-to-one balance of sounds to graphic markers. It was predominantly used for printing purposes. Pitman collaborated on phonotypy with Alexander John Ellis, with whom he also worked on the International Phonetic Alphabet. Phonotypy was, in fact, sometimes called the Pitman-Ellis alphabet. They experimented with as many as ten different variant font styles based on other modern (French, German) and ancient European (Old English, Greek) writing systems. Ultimately Pitman settled on new letters that hewed closely to the Roman.

Phonotypy (figure 5.2) was sufficiently proximate to everyday English that ordinary readers could parse it out, and indeed, using that appeal to

9. Kelly, "The 1847 Alphabet."

bring them on board was clearly a major factor in the conservative nature of the project as compared with the radical innovation of phonetic shorthand. It is absolutely essential that we recognize how very recent the standardizations of English typography were, such as the elimination of the long *s* and *ct* ligature as well as the discovery of the lack of uniform spelling: these small but decisive changes took place in the roughly fifty years between Webster's 1789 *Dissertations* and Pitman's publication of phonographic shorthand.

While shorthand consisted of a multiplicity of systems with an equally diverse assortment of names such as "brachygraphy," "tachygraphy," "stenography," "logography," and "phraseography," to name but a few, by the late nineteenth century they mostly shared a common basis in the phonetic-graphic interchange brought about by Pitman. We might plausibly say that the elimination of the hieroglyphic itself was the chimerical goal of shorthand and phonotypic reform. This is not to say that Pitman, Ellis, and others were tilting at windmills in the steadfast belief they could conquer all nonphonetic registers of meaning. They were well aware of the use of certain graphic characters from punctuation to universal symbols in mathematics, commerce, and so forth. It was, paradoxically, their very desire for an internally logical and utilitarian writing system that led to the ridicule and accusations of irrational behavior they endured in their own lifetimes and beyond.

In 1899 Mark Twain wrote an article entitled "A Simplified Alphabet," which explored the advantages and disadvantages of a simplified spelling system in comparison with Pitman's phonographic shorthand and its American imitators such as Eliza Boardman Burnz's *Phonic Shorthand*. Decrying how simplified spelling "sucked the thrill" out of familiar scripts, be they Greek, Hebrew, Russian, Arabic, or hieroglyphic, Twain came out in favor of shorthand's aesthetics and economy of expression. There was no need to buy a typewriter, after all, when one could fashion oneself into a human writing machine. Nonetheless, simplified spelling had its prominent supporters. Melville Dewey (1851–1931), the creator of the Dewey Decimal System that imposed the logic of standardization upon libraries throughout the United States, was a staunch supporter of simplified English who signed his name Melvil Dui.[10]

---

10.   Dewey also established the American Metric Bureau in Boston in 1876 to support national adoption of the metric standard.

As with Gutenberg before him, Pitman began with a shorthand transcription of the Bible to prove the merits of this new system of writing.[11] Unlike Twain, Pitman did not believe shorthand could truly supplant the alphabet—for that he provided the compromise measure of phonotypy— so he published various texts besides manuals and guides. Although we will explore in chapters 7 and 8 the shift from shorthand practice to the literary theories of mimetic capture that decisively contributed to the discourse of realism and the formation of the modern Japanese novel, a few observations about shorthand and literature in the Anglophone world merit attention. In the roughly sixty years from the invention of Pitman's phonography in 1837 to 1900, shorthand was practiced not only by Dickens; it was also incorporated into massively popular literature by Bram Stoker and Arthur Conan Doyle.

In a fitting connection with the forensics and secret codes of detective fiction, Arthur Conan Doyle's *The Sign of Four* and other novels had special editions published in shorthand.[12] Shorthand made its first narrative appearance in Conan Doyle's debut work, *A Study in Scarlet* (1887), with additional cameos in *The Memoirs of Sherlock Holmes* (1894) and *The Land of Mist* (1926). In *A Study in Scarlet*, the whodunit makes clear shorthand's centrality in the narratological circuit between fiction, journalism, and the police procedural:

> So thrilling had the man's narrative been, and his manner was so impressive that we had sat silent and absorbed. Even the professional detectives, *blasé* as they were in every detail of crime, appeared to be keenly interested in the man's story. When he finished we sat for some minutes in a stillness which was only broken by the scratching of Lestrade's pencil as he gave the finishing touches to his shorthand account.[13]

11.   As Albright notes, Pitman also produced portions of a Phonotypic Bible with Ellis, but the project ended (ironically enough) around Exodus due to disagreements over which type to use. See Albright's "The International Alphabet," 23.

12.   These texts may well have been published with the utilitarian purpose of increasing the speed and ease of comprehension for shorthand reporters, but it seems overly deterministic to conclude that shorthand-literate readers did not enjoy reading this script by which they plied their trade. As in other matters, it is often the case that contemporary scholars simply do not believe these experimental scripts, as I have called them, constituted a viable new form of written language.

13.   Conan Doyle, *A Study in Scarlet*, chap. 13.

The simultaneous production of a shorthand account lends a sense of immediacy to the confession that in turn bolsters the spectral presence of author-as-scribe and reader-as-observer in the room. It is equally implicit in the conduct of the journalist and police detective whose reports enter the crime in the public imagination and the crime ledger, respectively.

Shorthand also appears in the context of English language and spelling reform in George Bernard Shaw's *Pygmalion* (1916). The play itself, made famous by Henry Higgins' fastidious corrections of Eliza Doolittle's Cockney accent and social behaviors, was based upon the historical figure of amateur linguist Henry Sweet (1845–1912). An avowed competitor to Pitman, whose shorthand method he dismissed as "the pitfall system," Sweet was by all accounts an irascible man who played right into the caricature of the angry and self-righteous crusader. Yet in his choice of Sweet, Shaw reveals a keen understanding of the language and script reform scene of a generation earlier. It is worth quoting Shaw's preface to *Pygmalion* at length:

> English is not accessible even to Englishmen. The reformer England needs today is an energetic phonetic enthusiast: that is why I have made such a one the hero of a popular play. There have been heroes of that kind crying in the wilderness for many years past. When I became interested in the subject towards the end of the eighteen-seventies, Melville Bell was dead; but Alexander J. Ellis was still a living patriarch, with an impressive head always covered by a velvet skull cap, for which he would apologize to public meetings in a very courtly manner. He and Tito Pagliardini, another phonetic veteran, were men whom it was impossible to dislike. Henry Sweet, then a young man, lacked their sweetness of character: he was about as conciliatory to conventional mortals as Ibsen or Samuel Butler. . . . Those who knew him will recognize in my third act the allusion to the patent Shorthand in which he used to write postcards, and which may be acquired from a four and six-penny manual published by the Clarendon Press. The postcards which Mrs. Higgins describes are such as I have received from Sweet. I would decipher a sound which a cockney would represent by zerr, and a Frenchman by seu, and then write demanding with some heat what on earth it meant. Sweet, with boundless contempt for my stupidity, would reply that it not only meant but obviously was the word Result, as no other Word containing that sound, and capable of making sense with

the context, existed in any language spoken on earth. That less expert mortals should require fuller indications was beyond Sweet's patience. Therefore, though the whole point of his "Current Shorthand" is that it can express every sound in the language perfectly, vowels as well as consonants, and that your hand has to make no stroke except the easy and current ones with which you write m, n, and u, l, p, and q, scribbling them at whatever angle comes easiest to you, his unfortunate determination to make this remarkable and quite legible script serve also as a Shorthand reduced it in his own practice to the most inscrutable of cryptograms.[14]

Shaw thus surveyed the contemporaneous field of phonetic activists and found it wanting. His gentle mockery in the preface does not diminish his sincerity in calling Sweet the "hero" who inspired *Pygmalion*, a vital figure when a phonetic revolution in English seemed truly possible. We also cannot overlook the centrality to Shaw of shorthand as the notation system to fix the correct and standardized pronunciation of English.

It goes without saying that the dissemination of shorthand in Japan took place with even more far-reaching transformations of language and script. Having established some basic context for its reception by leading writers in the Anglophone world, we can now turn to its introduction and adaptation to Japan in the 1880s.

Hooked on Phonics

Credit for the creation of a Japanese system of shorthand is due to Takusari Kōki (1854–1938),[15] who adapted the Pitman-Graham system for use with the Japanese syllabary and convened the first school of instruction in shorthand on October 10, 1882, in Nihonbashi. He gathered a number of young students, who went on to establish their own shorthand systems. Shorthand also attracted considerable attention from national language reformers, educators, and government officials. Takusari initially called

14. Shaw, *Pygmalion*, 1–2.
15. Takusari was also known by the surname Minamoto.

his system of shorthand *bōchō kirokuhō* (hearing-record method), which he renamed about a month later as the equally voluminous *bōchō hikkihō* (hearing-notation method). These cumbersome titles were eclipsed in 1883 by political novelist and *Yūbin Hōchi Shimbun* editor Yano Ryūkei's more concise neologism *sokki* (literally, tachygraphy, or "fast-notation"). Although Takusari experimented with various names for his pioneering brand of shorthand, including endorsement of Yano's term, he referred back to Pitman's phonography as the precedent for Japanese shorthand, as we see in parallel English and kambun inscriptions on the title page of the manual *Shinshiki sokkijutsu* 新式速記術 (A new method of shorthand) from as late as 1904 (figure 5.3). Despite the twenty or more variations of shorthand in Meiji alone, in Japan as in the Anglophone context, proponents were similarly consistent in referring back to Pitman's phonography as the origin of Japanese shorthand.[16]

In addition to its relationship to phonography, which was still more closely affiliated with Pitman than Edison until the early twentieth century, Takusari's disciple Maruyama Heijirō (1865–1932) created a modified system in the early 1880s he called *kotoba no shashinhō*, literally, "a photographic method of words," or as I will refer to it hereafter, "verbal photography." The scene of writing that links orator and shorthand reporter into an inscriptive circuit is comically expressed in the cover art of Maruyama Heijirō's shorthand manual of the same name, *Kotoba no shashinhō* ことばの写真法 (Verbal photography, 1886). Blending the representational tropes of illustrated Tokugawa popular literature (*kana zōshi*) with iconography from the history of Western print culture (that is, scrolls and floral arrangements), it depicts a giant mouth-headed orator at the top of the picture plane addressing an audience of young men on his right, while a giant ear-headed shorthand reporter is busy at work on his left transcribing his every word (figure 5.4). The cover likewise prominently displays the cursive strokes of shorthand alongside the established kanji and kana scripts.

16. A practical exercise at the end of Maruyama's *Kotoba no shashinhō* teaches students that shorthand was invented by Pitman in 1837 and is known as "phonography" in English.

"*Phonography is the most Important, most Useful and most Noble employment of human*".——

速記術者人間職業中最緊要最有益最貴重者也

5.3 Title page to Takusari Kōki's *Shinshiki sokkijutsu* (New system of shorthand), 1904. Photograph courtesy of Waseda University Library.

5.4 Cover art for Maruyama Heijirō's *Kotoba no shashinhō* (Photographic method for words), 1886. Photograph courtesy of U.C. Berkeley Library.

Crucial to the dissemination of shorthand was not only the letters themselves, but their mode of notation. Maruyama's booklet contains the usual sound charts of kana and shorthand (figure 5.5). Yet if we look closely at the title at the top of the page in figure 5.5, we will notice a circular diagram like a pie in its center. This diagram reflects another universal trait of shorthand, the steady movement of the hand across the page. As seen in the sound chart that fills the page, phonetic values are captured in simple straight and curved lines, sometimes slightly looping, that do not require the hand to lift off the page. In so doing, shorthand turns the technical and aesthetic variability of calligraphy into a standardized operation.

Another shorthand system by one of Takusari's students was Hayashi Shigeatsu's *Hayagakitori no shikata* 早書き取りの仕方 (Fast notation, 1886). Whereas Takusari and Maruyama's systems were formulated in relation to kana, Hayashi linked his shorthand to Roman letters, albeit organized along the same phonological values as the kana sound chart. Hayashi enlisted Toyama Masakazu and Ōtsuki Fumihiko, luminaries from the Meiji educational bureaucracy, to pen its two forewords and fellow language reform advocate Terao Hisashi to write the postscript. Reflecting the respective camps of the Kana Society (Ōtsuki) and Romanization Society (Toyama, Terao, and Hayashi himself), *Fast Notation* features all of the major script reform movements, including shorthand, in a kind of reformers' omnibus project. The text's first preface was written in the Hepburn Romanized script, composed in classical grammar and signed "M. Toyama,"[17] by Toyama Masakazu, cofounder of the Romanization Society, coauthor of the first Japanese collection of free verse poetry, and future chancellor of Tokyo University. A strong supporter of Hayashi's shorthand, he also contributed a foreword to Hayashi's

---

17. More than a mere rhetorical flourish, this signature was a simple but powerful liberation from syllabary's consonant-vowel pair bond. It attacks the fundamental premise that Japanese spoken, and subsequently written, language is based on the so-called fifty-sound chart of syllabary. It is the legacy of this atomized letter that Sōseki manipulates in *Kokoro* (1914). Like the foreigner-gone-native who swims naked in and out of the beginning of the novel, the lone letter K, which coincides with the first initial of the title, is the anonymous reduction of the character who signifies modern alienation and its suicidal solution.

5.5 Sound chart for Maruyama Heijirō's *Kotoba no shashinhō*, 1886. Photograph courtesy of U.C. Berkeley Library.

Sokkijutsu daiyō 速記術大要 (An overview of shorthand), published in 1886. Toyama argues that shorthand has more to offer than merely a rendition of the standard dialect of the Tokyoite: it can also accurately capture the pronunciation of far-flung inhabitants of the country from Tōhoku in the north to Shikoku in the south. Unlike the restricted economy of kana, shorthand, he maintains, can discern between sounds such as *ha* and *fa*, which in Japanese share the same written sign, and articulate other sounds that might plausibly be verbally enunciated, but not recognized by conventional forms of Japanese writing. Thus, as provincial dialects (and in the ensuing decades, the foreign languages of colonized peoples) were increasingly coming under scrutiny as obstacles to the formation of a unified national language, shorthand held out the promise of putting them on an equal footing with the language of the capital:

Now shorthand is not only a code for expressing the speech of the solitary Tokyoite, but can represent people's speech from Sendai in the north to Tosa in the south: it should be thought of as a system that differentiates sounds such as *ja* and *dya*, *ha* and *fa*.

Sate, sokkijutsu nite wa hitori Tōkyōjin no on wo hyōsuru tame no kigō wo yōsuru nomi narazu, Nambu Sendai, Tosa Izumo no hito no on made wo mo hyōsuru koto no dekiru yō ni narite oraneba naranu ga yue ni, ja to dya, ha to fa tō no gotoki kubetsu wo ichi-ichi tateraretaru wa mottomo no koto to omowaruru nari.[18]

While based on alphabetic or syllabic systems, shorthand held out for its proponents the utopian ideal of transcending those limitations by introducing sound values otherwise not represented in mainstream discourse.

Consistent with the bodily discipline evidenced in other shorthand systems, Toyama emphasized the need to "train the ear" (*mimi no kyōiku*) as the first step toward the fidelity of recording. Hayashi, too, explicitly argues for shorthand as an embodied system of recording, whereby the hand is linked to the ear in a continuous circuit of faithful recording: "Fast-writing is the art of writing down exactly what the ear hears" (*hayagakitori wa hito no kotoba wo mimi ni kioku mama ni kakitoru no gakujutsu nari*).[19] Toyama reminds the reader that memorization alone is not sufficient to become a shorthand practitioner: "If one only learns the symbols but does not have a good ear, one will be unable to write down people's words just as they are" (*Kigō dake wo yoku shirite orite mo mimi ga warukereba, hito no iu koto wo sono mama ni kakitoru koto wa dekimaji*).[20]

In the Romanized afterword to *Fast Notation* (also composed in classical grammar), Terao Hisashi reiterates the political implications for

18. Toyama, "Jobun," in Hayashi, *Hayagakitori no shikata*, second unnumbered page. For the benefit of the reader I have refrained from italicizing the Romanized text in these longer passages.

19. Hayashi, *Hayagakitori no shikata*, 1. For convenience sake, I have used the standardized forms of kanji and kana; in the original text, Hayashi uses variant kana for *ni* and the nonsimplified form of *kiku*.

20. Toyama, "Jobun," in Hayashi, *Hayagakitori no shikata*, third unnumbered page.

capturing the breadth of spoken Japanese and making it available in writing to the nation. Moreover, in the perpetual game of one-upmanship that characterized competing shorthand systems, Terao praises Hayashi's new method as an improvement upon Maruyama Heijirō's photographic method and compares it to the nearly perfect transcription of oratory theatrical performances:

Rapid notation is essential to the Diet; without it the Diet cannot satisfactorily shine its light [upon the nation]. The country is vast and the people many, yet only a scant few may sit in attendance of its proceedings. It is therefore of the utmost importance that the speeches of the members of the Diet be written down *just as they are* delivered and made available to the public. . . . For instance, typical transcriptions are like a badly drawn caricature, while the rapid notation of Mr. Hayashi and his colleagues are like a photograph [*shashin no gotoshi*]: not an ordinary photography, but like one of the fast-developing kind taken by Ezaki Reiji. Or if we were to liken it to a play, the usual method of writing down oratory can be very good, akin to reading the scenarios in *Kabuki Shimpō*; this new method is better than reading the *jōruri* [puppet theater] scripts, it is like listening to the recitations of Ayase or Aioi: of course, while it cannot compare with watching a play it isn't terribly hard to imagine, you feel like crying yourself during the sad parts, at the tragic parts you grind your teeth in frustration, all of the things aficionados of *jōruri* can appreciate.

Kokkwai ni hayagakitori wa tsukimono nari; hayagakitori nakute wa kokkwai mo jūbun ni hikari wo hanatsu koto atawarazu-beshi. Chihō wa hiroshi, jimmin wo ōshi, kokkwai no bōchō ni deru mono wa wazuka no ninzu nareba; zehi daigishi no enzetsu nado wo *sono tōri ni* hikki shite yo ni ōyake ni suru mono nakaru-bekarazaru wa mochiron nari. Jūrai no hikki wa tatoeba heta no kaita nigao no gotoku, Hayashikun nado no hayagakitori wa shashin no gotoshi; shashin mo futsū no shashin de wa naku Ezaki Reiji no hayautsushi no gotoshi. Mata shibai ni tatoyureba, shibai ni shikata nite kakitoritaru enzetsu wo yomu wa, yohodo yoi no de, *Kabuki Shimpō* no sujigaki wo yomu ga gotoku; kondo no shikata no wa jōruribon wo yomu yori mo yoku, ōkata wa Ayase ka Aioi no jōruri wo kiku ga gotoshi: mochiron shibai wo miru hodo ni wa nakeredomo, hidoku sōzō wo tsuiyasazu shite, kanashii tokoro wo onozukara nakitaku nari, kuchioshii tokoro wa onozukara sesshi-yakuwan shitaku naru koto, jōruri wo konomu mono no mina shiru tokoro nari.[21]

Ezaki Reiji (1855–1910) was the first in Japan to produce dry plate photography in 1885, which he called *hayatori shashin*, and which surpassed the slower developing glass plate technology. Hayashi's *hayagakitori no shikata*, a Japanese equivalent to the Sinologism *sokkihō*, was clearly calculated to resonate with Ezaki's breakthrough in photographic technology. Trumping Maruyama's emphasis on the common virtues of photography, Terao rode the coat-tails of Hayashi's claims for the speed and effectiveness of his own new-and-improved system.[22]

Kamei Hideo dismisses the claims of *Fast Notation* and related texts in *Meiji bungaku-shi* (Meiji literary history, 2000), remarking that even if shorthand could capture regional accents (*okuni-namari*) and speech mannerisms (*kotoba-guse*), such a "hyper-correct" style would simply be too difficult to read. It is a remarkable admonition considering the stakes for preserving local, and indeed, minority voices in the early Meiji political process. What Kamei finds objectionable is the unreaderliness of the text in the Barthian sense, or what we might plausibly call "dyslexia," rather than the failure of the text itself to make sense. By impeding or rendering opaque the transparent surface of meaning, composition from shorthand threatens a breakdown in the pleasurable act of reading and forces the reader to confront the materiality of the text. But there is a secondary, and far more serious, problem with Kamei's objection, which is the reinforcement of a standardized milieu that secures at any cost the *aesthetic* operability of the imagined community. Kamei appears willing to endorse the unified style for its aesthetic effects at the expense of its

---

21.   Hayashi, *Hayagakitori no shikata*, 33–34 (emphasis in original).

22.   Contrary to Toyama's wishes for shorthand to preserve dialectical variation, botanist Yatabe Ryōkichi (1851–1899), one of Toyama's collaborators on *Shintaishi-shō* and fellow member of the Romanization Society, published *Romaji Hayamanabi* (Fast learning of roman letters, 1886) through the Maruzen bookstore on the basis of its capture of standard dialect. It, too, offers its readers the advantages of a quick and easy learning method, only with the unabashed pitch that it was "based upon the standard pronunciation of Tokyoites who received normal school education" (*jinjyō no kyōiku wo uketaru Tokyojin no aida ni okonowaruru hatsuon wo motte narubekitake hyōjyun to suru koto* [1]).

ethical costs: silencing in form and substance the dissenting voices enabled by shorthand. In fact when the government began to police politically "licentious" speech, the People's Rights advocates sought to evade the censors through *rakugo-* and *kōdan-*style performances, whose shorthand transcription became a more covert means of disseminating revolutionary democratic and socialist ideas.[23]

Contemporaneous with these and other burgeoning developments in shorthand, Yano Ryūkei included shorthand in his historical and critical explorations of literary styles and scripts. Leaving aside for the moment his collaborations with Wakabayashi Kanzō to transcribe his political novel *Keikoku bidan* 経国美談 (Illustrious statesmen of Thebes, 1883–1884) and his commentaries about the future of Japanese writing in its two prefaces, which I will explore at length in chapter 7, Yano's treatise *Nihon buntai moji shinron* 日本文体文字新論 (A new theory of Japanese style and script, 1886) was published by Maejima Hisoka's *Yūbin Hōchi Shimbun*, where Yano was the editor and employed several shorthand reporters, including Wakabayashi Kanzō and Sakai Shōzō. The treatise was a boldly comparative analysis of the new possibilities for stylistic and script reform across a span of modern and ancient languages. I want to bring into relief several key aspects of Yano's study relating to shorthand and surmounting the incommensurability of speech and writing.

In a familiar criticism of the unfiltered transcription of speech, Yano compares the polished style of a professional theatrical storyteller (*kōdanshi*) and "the average person" (*jinjō no hito*) in order to point out the need for a concise yet lively style. He reflects on the need to create an "ordinary language" (*nichijō no kotoba*; *jōgo* for short) that would enable everyone to become what we might call speech-writers: "[It is] a serious issue that we must produce an ordinary language in Japan that can be written just as it is [spoken]" (*Nihon no jōgo wo sono mama ni bunshō to nashi ebeki ya ina ya ichi dai mondai* [*nari*]).[24] The speech one writes should not be disorganized or excessively prolix, but should be normalized

---

23. See also Komori's *Nihongo no kindai*, 45 and 56–60.

24. Yano, *Nihon buntai moji shinron*, 43. While premodern thought and literature are rife with examples of the so-called "common man" (*bonbu*), Yano's reference to the average person and common language is indelibly linked to the establishment of normal schools (*jinjō gakkō*) and other standardizing institutions and measures.

through the oratory arts (*wagei*). Yano repeatedly alludes to the figure of the storyteller in *rakugo* and *kōdan* in his discussion of a written vernacular, although he eschews the term *genbun itchi* popularly attributed to Mozume Takami. In a continuation of this argument, Yano holds up equivalent passages of the literary (*bungotai*) and spoken (*kōgotai*) forms of English to demonstrate the problem is not unique to Japanese alone.

Yano devotes the final chapters to exploring the comparative advantages of phonetic (*onji*) and figural (*keiji*) scripts. Unlike shorthand practitioners with their concerns for the hand and ear in recording speech, he primarily investigates practices of reading and the mechanics of the gaze. Interlinear sentence spacing, the composition of letters, and the movement of the reader's eyes are all crucial factors in assessing the visual field he calls "the world of the eye" (*me no sekai*). Yano was concerned not only with the popularization of typography, but with its physiological demands on the eye. Yano describes the mechanical workings of the human eye as virtually equivalent to a photographic lens (*shashin kikai no megane*).[25] In a series of anatomical illustrations, the eye is dissected into optical nerves and muscle fibers, followed by geometric charts mapping the angles of movement for the eyes and resultant potential for eye strain. He compares different conventions in reading order in Chinese and Japanese (from top to bottom and right to left), English (from left to right and top to bottom), and alternatives from ancient Greek, Sanskrit, Arabic, and Mongolian. Yano argues that the reading of Chinese characters in fact induces less strain than Western languages, whose arrangement on the page induces an asymmetrical movement in the left eye relative to the right. Such was the seriousness with which the fitness of Japanese writing on the world stage was to be evaluated.

Yano disapproved of simplistic dichotomies between phonetic and figural writing systems. Throughout the text he is at pains to point out all sorts of anomalies such as the fact that even in the modern West there are nonverbal symbols used all the time such as $ and &.[26] In a further rebuttal that strongly suggests a connection with Nishi's original notes to his *Ency-*

---

25. Ibid., 142.

26. Although Yano does not mention it, the ampersand (&) in fact originated in the Notae Tironianae as a ligature of the Latin "et." In this sense, its provenance is more closely linked to the antecedents of shorthand than he may have known.

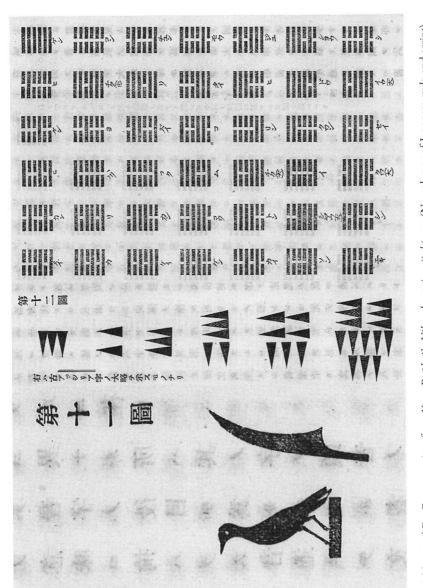

5.6 Near and Far Eastern scripts from Yano Ryūkei's *Nihon buntai moji shinron* (New theory of Japanese style and script), 1886. Photograph courtesy of University of Chicago Library.

clopedia, Yano situates ancient Phoenician and Assyrian script alongside the Chinese *I-Ching* for the purpose of demonstrating the existence of figural scripts in Western antiquity and phonetics in Chinese antiquity (figure 5.6). While he let slip the opportunity to comment upon the phonetic aspects of Chinese characters, Yano sought to transcend the binary opposition of phonetically standardized and therefore supposedly superior languages of the West and the heterogeneous and figural makeup of Japanese.

After extensively surveying ancient and modern scripts, Yano comes around to shorthand in the chapter on reading order. Referring to what he calls a "radical view" amongst Western language reformers (*chokushin no rikutsu*, or "forthright logic," is parenthetically glossed in katakana *radikaru bīu*), he reminds the reader that the alphabet is not a recent invention such as the steam engine or telegraph, but a legacy of the ancient Phoenicians that has been passed down to the present day with incomplete standardization. If one were to look for a contemporary analog to those more recent inventions, he suggests, it would be the invention of phonetic shorthand. Indeed, Yano allows that if certain trends continue, shorthand may yet displace the alphabet in the West. He insists that

> In addition to their form, the letters of shorthand notation are easy to read. Lately, in Europe as well, shorthand has gained considerable force, with the number of young businessmen who can read it on the rise. There are even arguments in favor of replacing the alphabet with these letters. If we are going to reform Japanese writing, we might be better off forgoing old-fashioned Romanization and go straight ahead to shorthand instead.[27]

> Mata katachi no ue yori iu mo sokkihō no ji wa miwake yasushi kinrai nite wa Ōshū nite mo sokkihō ōini ikioi o ete wakaki shōnin nado wa kore o yomieru mono zōka suru ni itaru kōrai wa kore no ji o motte Rōmaji ni kawaru yo no naka to suru beshi to no hyōban mo aru gurai ni itareri sareba? Nihon no moji o kaeru hodo naraba kyūbutsu no Rōmaji yori mushiro chokushin shite koko ni itaru koso yoroshi karubeshi.

Although this can hardly be misconstrued as a wholesale endorsement of shorthand, as early as 1883–1884, Yano introduced shorthand in the production of his political novel *Illustrious Statesmen of Thebes* and regarded it as integral to the formation of a modern language and literature prior to

27. Yano, *Nihon buntai moji shinron*, 209.

the transcription of Enchō's *The Peony Lantern*. If shorthand was not yet the definitive answer to the establishment of a modern Japanese language and script, it was articulated as a contender whose merits lay not in the accumulated weight of tradition, but in their unencumbered utility.

Another commentator in the heady days when shorthand seemed poised to effect a major change in Japanese letters was Basil Hall Chamberlain (1850–1935), the first professor of linguistics at Tokyo Imperial University. Written only six years apart, Chamberlain's *A Handbook of Colloquial Japanese* (1889) and *Practical Introduction to the Study of Japanese Writing* (1905) nevertheless stake out diametrically opposing views of shorthand and experimental scripts. While the *Handbook* includes a detailed description of the transcription of rakugo and political oratory as a new cultural phenomenon, and prescribes its use as a pedagogical tool for foreigners studying Japanese, the *Practical Introduction* traces the arc of shorthand from bright promise as an instrument of linguistic change to bitter disappointment. From the late 1880s, Chamberlain was quick to note the disparity between spoken and written languages right on the cusp of the unified style's ascension. As indicated by its title,[28] the *Handbook* sets aside the complexities of the formal written language and concentrates instead upon mastery of the spoken. It accomplishes this in part by relying on one of the most popular modern Japanese "literary" texts: Enchō's *The Peony Lantern*. The *Handbook* features the first two chapters of *The Peony Lantern* transliterated into Romanization side by side with its English translation. In the process of explaining his choice of teaching materials, Chamberlain expresses the commonly held view about the incommensurability of Japanese written and spoken language:

> The Japanese . . . do not write as they speak, but use an antiquated and indeed partly artificial dialect whenever they put pen to paper. This is the so-called "Written Language." Of the few books published in the Colloquial, the best are the novels of a living author named Enchō. The student, who does not wish to trouble about the [Chinese] characters, cannot do better than write out one of these books from his teacher's dictation.[29]

28. The text was truly internationalist in its distribution, as it was carried simultaneously by the publishers Trübner & Co. in London; Hakubunsha in Tokyo; and Kelly & Walsh Ltd. in Yokohama, Shanghai, Hong Kong, and Singapore.
29. Chamberlain, *Handbook of Colloquial Japanese*, 10.

This characterization of Enchō as a novelist throws open some very familiar assumptions about the settled definition of Japanese *and* English literature at the cusp of its transformation in the mid-1880s. The idea of Enchō as author did not originate with Chamberlain either. As we will see in the next chapter, the definitions of author, text, and audience were somewhat different in the shorthand transcribed literature of the 1880s before literary theory by Tsubouchi Shōyō and others established what we think of today as the canonical order. Chamberlain's preface also advises the Sinographically illiterate foreigner to learn spoken Japanese by rote imitation of shorthand transcriptions.

> Occasionally too, the newspapers, the "Transactions" of the Education, Geographical, and other learned Societies, and such collections of lectures and speeches as the *"Taika Ronshu,"* the *"Kodan Enzetsu-shu,"* etc., print a lecture exactly as it was taken down by the short-hand reporter from the mouth of the lecturer, though the more usual practice is to dress everything up in the Written Style before it is allowed to appear in print.[30]

Chamberlain commits a subtle misrepresentation here. Clearly, he has in mind the catchphrase *ari no mama*, but means to say the shorthand has already been rewritten into a more accessible version in kana or Roman letters. Still, it was a convenient way for students of the language to read a transcription of spoken Japanese before it was transformed beyond recognition into a formal written style.

As a leading member of the Romanization Society, Chamberlain was a strong advocate for linking Romanization to the unified style in the late 1880s. Yet he cautioned against a mere substitution of scripts without concordant stylist reforms. As Nanette Twine summarizes,

> On 19 March 1887, he gave a lecture titled "Gembun'itchi" [*sic*] at the second general meeting of the [Rōmaji] Club, the text of which was published in colloquial style in *Rōmaji Journal* in May. . . . [He argued that] to take away the visual cues offered by Chinese characters was merely to compound the difficulty of written Japanese. It was no use replacing Chinese words with classic Japanese expressions; the archaic language simply

30. Ibid., 11.

did not have the vocabulary to express modern concepts. The solution, he believed, lay in using colloquial style. . . . Partly in response to these theories and partly in response to the growing influence of the *gembun'itchi* movement in society at large, there did in fact occur a change from *kambun kuzushi* [modified Chinese for Japanese readers] to colloquial style in the essays published in *Rōmaji Journal* after April 1887.[31]

Chamberlain undoubtedly saw shorthand in the late 1880s as the key to achieving both Romanization and unified style reforms. Membership in the Romanization Society was close to 6,700 at the time of Chamberlain's lecture, up from around three thousand in June 1885, and it swelled to about ten thousand by the end of 1888.[32] For reasons that have not been adequately explained, however, it suffered a rapid change in fortunes in what Twine calls "the backlash against over-enthusiastic adoption of western customs"[33] and was dissolved in 1892. There would be no major revival until Tanakadate Aikitsu's establishment of the Nippon-no-Romaji-Sya in 1909.

Skeptical about the inaccessibility of Romanization that was divorced from the context of colloquial style, Chamberlain had already voiced fears in his 1887 lecture/article that the good ship "Romanization Maru" would surely be dashed on the rocks and sink (*anshō ni "Romanization Maru" ga tsuki-atari, tachimachi chimbotsu shi*)[34] if the Romanization Society did not take steps toward embracing the unified style. Chamberlain appears to have experienced profound disillusionment in the intervening years and made a reversal of his earlier sanguine advice. Occasioned by the failure of Romanization and kana to make the slightest dent in the hegemony of the mixed script, or what he calls "a backbone of Chinese characters with Kana ligaments,"[35] Chamberlain cautions students in his *Practical Introduction to the Study of Japanese Writing*

31. Twine, *Language and the Modern State*, 243–44. For a reproduction of the text in Japanese as it appeared in the *Rōmaji zasshi*, issue 24 (April 10, 1887), see Yamamoto's *Kindai buntai keisei shiryō shūsei*, 358–60.

32. Twine, *Language and the Modern State*, 240.

33. Ibid., 244.

34. Chamberlain, "Gem-bun Itchi," in Yamamoto, *Kindai buntai keisei shiryō shūsei*, 360.

35. Chamberlain, *Practical Introduction to the Study of Japanese Writing*, 5.

about the inadequacy of learning only kana or *Rōmaji*. He became exquisitely contemptuous of both campaigns, dismissing them as no better than the lunatic fringe of Pitman's phonotypy:

> And do not come and tell us—as if they constituted some startling new factor about to revolutionise Japan—of booklets in Kana or in Roman, which you have lighted upon in some nook or corner. Such things exist,— have long existed: but they possess the same importance (or unimportance) as the "Fonetik Nuz," or those English treatises on "Little Mary and her Lamb" and cognate topics which sometimes drip from the press in words of one syllable exclusively.[36]

It is precisely this condescending attitude and out-of-hand rejection of experimental scripts that has been reproduced by canonical Japanese (and, for that matter, English literary) studies for the better part of a century; namely, that they were marginal activities unworthy of further investigation. We must, however, distinguish between Chamberlain's stance as a contemporary participant and one-time advocate of those scripts on the one hand, and the erasure of these material and media histories particularly from postwar scholarship, on the other. Yet as we will see in Isawa Shūji's imperialist applications of Alexander Melville Bell's Visible Speech in the next chapter, and the theories of transcriptive realism profoundly shaped by the media concepts and compositional imperative of shorthand in chapters 7 and 8, the legacy of shorthand pervaded discourses of Japanese modernity well beyond the obscurantism mocked in this passage. Moreover, as Kittler reminds us, Edison's first composition on the phonograph was none other than "Mary Had a Little Lamb," bellowed at full volume to compensate for his own partial deafness and the lack of an amplifier.[37] The phonograph, of course, was itself conceptually piggybacked onto Pitman's manual phonography. Prior to the advent of other machines of recording and transmitting such as the telephone invented by Alexander Graham Bell—the first "native speaker" of Visible Speech— media capture of the voice depended on phonetic scripts captured by circuits of human hands, ears, and eyes.

36. Ibid., 6.
37. Kittler, *Gramophone, Film, Typewriter,* 21.

CHAPTER 6

Parsing Visible Speech

This chapter examines the experimental phonetic script known as "Visible Speech," developed by Alexander Melville Bell in the late 1860s and adapted by Isawa Shūji (1851–1917)[1] for Japanese national and colonial education between the late 1870s and 1910s. From its inception in English, Visible Speech was intended as an aid to the deaf, stammerers, and other verbally disabled persons. Created in advance of the International Phonetic Alphabet (IPA), Visible Speech sought to restore proper enunciation to the speech impaired through a new graphic script that Bell believed literally mapped onto the parts of the mouth, throat, and vocal cords. In an era of manual recording and transmission prior to mechanical devices such as the telephone that his son invented, Bell averred that Visible Speech provided the most phonetically and phonographically precise script the world had ever known. Arguably, however, it was not Melville Bell or Graham Bell, but Isawa who disseminated it most widely and came closest to applying it on any significant scale. In its adaptation, he further demonstrated the extent to which experimental scripts that originated in the Anglophone world were integrated into the ambit of Japanese modernity.

---

1. Isawa romanized his name as Isawa Shuje, which is also sometimes written as Izawa. I have followed contemporary usage.

Alexander Melville Bell and the Human Speaking Machine

Visible Speech was unquestionably Melville Bell's major life's work, but it was not his only effort at script reform. In the late 1880s, he created a modified alphabet called "World-English,"which promised to be more of a compromise script when Visible Speech failed to take hold on a large scale. In an appeal to the printing industry on both sides of the Atlantic, Bell dedicated Visible Speech and World-English to "Conductors of the Press [who] have the power of greatly facilitating the object of this work, by making it known; or of retarding it, by simply ignoring the effort. Opposition is not to be looked for from any quarter."[2] He also wrote shorthand manuals such as *Popular Shorthand or Steno-Phonography* (1892). Although he admitted that Visible Speech could not compete with shorthand in terms of speed, Bell maintained its phonetic accuracy superseded Pitman's and other shorthand methods. In this way, Bell located his work on script reform squarely in the commercial and practical domain, as well as the utopian.

Notwithstanding his considerable stature in nineteenth-century language and script reform, Melville Bell has been largely forgotten by modern linguistics. Of course he is hardly alone in this respect—Isaac Pitman, Alexander John Ellis, and other prominent phoneticists, orthographers, and artificial language planners of the same era have similarly been marginalized. My concern here is less for Melville Bell's restoration to glory, however, than to provide the necessary historical context to understand how and why Isawa embraced this script with the enthusiasm of a religious convert, and in turn wanted to use it to advance Japan's own imperial policies in the home islands, Taiwan, Korea, Manchuria, and elsewhere.

There is an implicit yet widely shared conceit that English and a handful of other major European languages became fixed in their modern forms by the mid-nineteenth century, and hence it was only the late modernizing states such as Japan that had to deal with language reform. Benedict Anderson's *Imagined Communities* differentiates between waves

2.   Alexander Melville Bell, Prologue, *World-English*.

of modern nationalism in part on the basis of this linguistic criterion, and thus between countries already unified by language and those for which it constituted a major obstacle. While this is an attractive idea, it does not exactly cleave to the historical record. In the nineteenth century, British and American intellectuals time and again betrayed their anxieties over the inadequacies of the English alphabet and incommensurability of speech and writing. These concerns, and those of language reformers throughout Western Europe, were tempered by a growing sense of the possibility of new techniques and technologies of writing to overcome old limits such as deafness and mutism. It further held out the utopian promise of unifying the many disparate linguistic communities that were now coming into contact in the heyday of European and American colonial expansion.

In order to demonstrate the efficacy of Visible Speech, Bell taught it to his two young sons, Edward (1848–1867) and Alexander Graham Bell. So thoroughly were they trained in the system that their father would confidently put their skills to the test in public spectacles. It was Bell's contention that Visible Speech enabled its adepts to accurately record and transmit virtually any sound. The Bell brothers would thus be put before rooms full of skeptics and asked to transcribe whatever their audience could come up with in English or foreign languages or even animal sounds. This bombastic, P. T. Barnum-esque publicity campaign for Visible Speech is reflected in the florid wording of the title page of the book by the same name (1867), which brings to mind the signboards of circus and vaudeville sideshows (figure 6.1).

Unlike shorthand, which was still closely based upon alphabetic or syllabic sound values, Visible Speech was composed of what he called "ten radical symbols" that are combined to form consonants, vowels, glides, modifiers, and tones. As its name suggests, Bell's universal alphabet has a pictographic component that locates sounds according to their physiognomic placement in the human vocal apparatus. As Bell explains, his goal was "to construct a Scheme of Symbols, which should embody the whole classification of sounds, and make each element of speech shew [*sic*] in its symbol the position of its sound in the organic scale."[3] In this

3. Bell, *Visible Speech*, 18.

ꟿ|ꟿ|ꟼꟽ| ꞇꝺ||ꝺꟽ:

VISIBLE SPEECH:

THE SCIENCE OF

ꝺꞇꟽ|ꟻ|ꝩꟿ|ꟽ |ꟽꟿ|ꟼ|ꝺ|ꝺꟽ:

UNIVERSAL ALPHABETICS;

OR

SELF-INTERPRETING PHYSIOLOGICAL LETTERS,

FOR THE WRITING OF

ALL LANGUAGES IN ONE ALPHABET.

ILLUSTRATED BY

TABLES, DIAGRAMS, AND EXAMPLES.

BY

ALEX. MELVILLE BELL, F.E.I.S, F.R.S.S.A.,

PROFESSOR OF VOCAL PHYSIOLOGY,
LECTURER ON ELOCUTION IN UNIVERSITY COLLEGE, LONDON,
AUTHOR OF 'PRINCIPLES OF SPEECH AND CURE OF STAMMERING,' 'ELOCUTIONARY MANUAL,'
'STANDARD ELOCUTIONIST,' 'EMPHASIZED LITURGY,'
'REPORTER'S MANUAL,' &C., &C.

INAUGURAL EDITION.

SIMPKIN, MARSHALL & CO.: LONDON;

N. TRÜBNER & CO.: LONDON AND NEW YORK.

SOLD BY ALL BOOKSELLERS.

1867.

Price Fifteen Shillings.

6.1 Title Page to Alexander Melville Bell's *Visible Speech*, 1867. Photograph by the author.

respect it is similar to *hangŭl*, which is also to some extent derived from signifiers for the mouth and other parts of Chinese characters.[4]

Visible Speech opens with a passage describing the human being as biological machinery, with the integrity of the human body made over into a set of interlocking parts. Its premise is consistent with Kittler's observations that deficiencies that interfere with the seamless operations of mind and body paradoxically help to reveal their inner workings. For Bell phonocentrism was replaced with physiological dissection, which was then translated into graphic notation:

> The lungs constitute the bellows of the speaking machine; the larynx, the pharynx, the soft palate, the nose, and the mouth, modify the breath into elementary sounds of speech.
>
> The lungs are enclosed within the chest, and in healthful respiration they are acted on chiefly by upward pressure of the diaphragm, or midriff, which separates the chest from the abdomen. In faulty respiration the sides of chest are drawn in upon the lungs to force out the breath, and the natural action of the diaphragm is reversed. Stammerers generally exemplify this error.
>
> The breath driven from the lungs ascends the windpipe, and its emission is rendered audible only by the resistance it meets with in the throat, the nostrils, or the mouth.
>
> On the top of the windpipe is placed the larynx,—the vocalizing part of the speaking machine. The larynx is practically a box, the cavity of which is susceptible of a multitude of modifications affecting the pitch of the voice. The orifice of the larynx—the glottis—may be perfectly closed, fully expanded, or contracted in any degree.
>
> When the gutteral passage is fully expanded, the passing breath creates no sound; but when the glottis, or aperture of the larynx, is definitely narrowed, its edges vibrate and produce the sound which is called voice. Voice is thus the mechanical result of vibration along the edges of the glottis.[5]

This brief description of the action of breath, voice and speech is supplemented by a diagram showing a cross-section of the human mouth, nose,

---

4. In Chinese and Japanese the character for the mouth, 口, functions as a phonetic radical (reproduced in smaller size alongside other ideographic components) and as an ideogram in its own right.

5. Bell, *Visible Speech*, 11.

and throat. It is a display of the body as living machine. In lieu of the metaphysical and philosophical musings about spirit, breath, and voice are the mechanistically determined origins of speech. Moreover, for Bell the phonogram could be improved to pictographically represent those schematized parts of the vocal apparatus. As a result, where Pitman and subsequent shorthand inventors sought to recast and expand the range of phonetic values as straight and curved lines for the purposes of fast, accurate recording, *Visible Speech* offered a physiological basis for its new script (figure 6.2). Visible Speech was supposed to be capable of recording not only human language, but virtually any sound from animal vocalization to mechanical noise. In advance of the phonograph, Visible Speech was to enable the human ear-to-hand circuit to utilize this script as the perfect recording instrument. (figure 6.3) In *Optical Media*, Kittler observes, albeit without mentioning Melville Bell or necessarily showing awareness of his achievements, that "phonographic recording, which at that time was also called visible speech, became the accepted thing among European physiologists." These physiologists proceeded to invent a host of machines for registering the disparate phenomena of human kinesiology and animal locomotion, as Kittler observes: "At the Collège de France, one device emerged after another: a heart recorder, a pulse recorder, and finally also a device that was connected to the four extremities of animals and could record their movements. None of these devices bore the least similarity to photographic cameras, but rather they worked, exactly like visible speech, with a pencil and a steadily moving paper cylinder."[6] What Kittler identifies as the common thread of Visible Speech and these other "visible speech" devices is the attempt to make previously inchoate aspects of human and animal existence both readable and writable—that is to say, subject to the principle of standing-reserve.[7]

Melville Bell envisioned his system as solving the problems of the deaf, dumb, blind, speech-impaired, and illiterate, as well as providing a means for accurate transcription of foreign languages, applications for the telegraph, and the establishment of unwavering pronunciation standards. Last but not least, he held out enticing benefits for anthropologists, missionaries, and imperialists to preserve dying dialects and to trace what

6.   Kittler, *Optical Media*, 155.
7.   Ibid., 155.

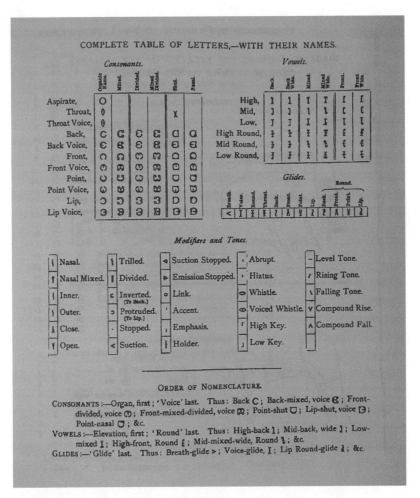

6.2 "Complete Table of Letters" sound chart from *Visible Speech*. Photograph by the author.

he called the "AFFINITIES OF WORDS" in order to enable "speedy diffusion of the language of a mother country throughout the most widely separated COLONIES." Most portentously, he envisioned

the world-wide communication of any specific sounds with absolute uniformity; and consequently, the possible construction and establishment of a UNIVERSAL LANGUAGE. . . . By means of Visible Speech, if at all,

INTERJECTIONAL EXERCISES
ON THE RUDIMENTAL CONSONANT SYMBOLS.

Cʃ	disgust.	ᘓ·	quiet sneering.	ᘎ()ᘎ	impatience.
Cʃʃ	snarling	ᘓ·	"	ᘎɔD·	spitting.
Cʃ<	snoring.	ᘎ·	"	()ɔᗝ	blowing from point of
Cʃʃ<	"	ᗝ·	"		tongue.
Cʃʃ	hawking.	ᗝ<ᗝ<	sniffing.	Dᵒᘎ<	sucking.
Cʃʃ	gargling.	ᗝʃ<	examination of odour.	ᘎᗏᗏ◄	tasting.
ᘓʃ	hissing.	ᘓᘎ	ridicule.	ᘎ◄ ᘎ◄ ᘎ◄	vexation.
ᘒʃ	hushing.	ᘓCʃ	"	ᘎʃ◄ ᘎʃ◄	inciting.
ᘒʃ	hurrying.	ᘓC	"	ᘎ'ᑊ► ᘎ'ᑊ►	"
ᗝᘒ	silencing.	ᘓCʃ	"	D◄	kissing.
ᘓʌʃ	blowing to cool.	ᘓʃ	a suppressed chuckle.	Dᗝʌʃ◄	chirping.
ᗝ<	sipping.	ᘓʃ	" "	ᘓʃ·O'ᘎʌʃᘒ	sneezing.
ᗝʏʃ	faintness from heat.	ᘎʃ	" "	ᘒᑊᗝ◄	a flap of the tongue.
ᗝʃ	a semi-whistle.	Dʃ	" "	ᘎᘒᑊᗝ◄	a clicking flap.
ᗝᗝᗝ	incredulity.	ᘓᘓʃʃ	snickering.	ᘎᘒᑊᗝᵒᗝ◄	" like the
ᗝᘓᗝ	"	ᘓᘓʃʃ	"		gurgle of decanted liquid.
ᗱᘓʃ	distaste.	ᘎᘎʃʃ	"	ᘓᘒʃʃ	the cry of a quail.
ᘒ<	pain.	DDʃʃ	"	ᘓᘒʃʌʃ	the grunt of a pig.
ᗘ<	"	ᘎᘓ	annoyance.	ᗝ()ʃᵒᗝʃ	the whirr of a partridge.
ᗱ<	"	ᘎᘓ	incredulity.	ᘓʃᑊᘓʃᘓ ᘓʃᑊᘓʃᘓ ᘓʃᑊᘓʃᘓ	
ᘒ<	"	Dᘓ	"		the sound of a grinding wheel.
ᘒ<	acute pain.	ᘎᘒ	contempt.	CʃᘓCʃᘓʃᑊᘓᘓ·	the sound of
ᘓ<	"	Dᘒ	"		planing wood.
		Dᘓ	abhorrence.		
		Dᘒ	"		

Cʃʃ< Cʃᑊ Cʃ< Cʃᑊ Cʃʃ< Cʃᑊ ᘓʃ< ᘓᘓʃᘓ· the sound of sawing wood.

6.3 "Interjectional Exercises" sound chart from *Visible Speech*. Photograph by the author.

this Dream of Philosophers will be realized. The foundation is laid, and the Linguistic Temple of Human Unity may at some time, however distant the day, be raised upon the earth.[8]

Bell's architectural vision amounted to a restoration of the prelapsarian Tower of Babel, where all humanity was unified under a common tongue. The supreme rationality of the Victorian linguistic scientist came with a twist: the triumph of a true phonetics over hieroglyphics.

When Visible Speech failed to gather a widespread following in the ensuing two decades, Bell revised the dominion of his temple to "World-English," an essentially alphabetic script akin to Pitman's phonotypy, which sought to improve upon existing letters, rather than replace them outright. Bell further appealed to the distinction between "Literary English" and "World-English" to avoid conflict with traditionalists and conservatives. In an echo of Mori Arinori's simplified English, he insisted it was the duty of government to promulgate this "amended scheme of letters" in public schools and other social agencies. For the foreseeable future, the gulf between written and spoken language would persist unabated. English would continue to grow as a "mother tongue" to millions of speakers, while "spelling must remain a separate art, pictorial in nature, and chiefly learned by the eye."[9] Bell attacked the arguments for planned languages such as Volapük, arguing that English speakers needed only orthographic improvements to enable their own language to fulfill its destiny, and to prevent a debased, pidgin language from usurping its place:

The English language has been, itself, steadily reaching out towards universality. It has covered the North American continent and the islands of the antipodes. It has become a necessity wherever English or American navigators penetrate. India, China and Japan are teaching it in their schools. Commerce has invented a barbarous variety of it as a Port-language, called "Pigeon-English" [*sic*]; and, but for a want of an explicit system of letters, it would, long ere this, have fully filled its destined place.[10]

8. Bell, *Visible Speech*, 21 (emphasis in original).
9. Bell, *World-English*, 27.
10. Ibid., 25.

Of course, Alexander Melville Bell met with only modest success amongst teachers of the deaf, and his son's similar efforts with Visible Speech would in time be dramatically overshadowed by his invention of the telephone. Ultimately, the legacy of Bell's experimental script was not a worldwide web of English, but its promulgation in the service of national and imperial Japanese on the archipelago and in East Asia, courtesy of Isawa Shūji.

Isawa Shūji and Imperial Linguistics

Despite the considerable attention given to Isawa Shūji in recent postcolonial scholarship by Oguma Eiji, Komagome Takeshi, Shi Gang, and others, to date there has been virtually no mention of Visible Speech in Isawa's thought on colonial and national education, much less its place in the broader discursive transformations of language and script reform. It is fair to say that in Japanese, as in English, Visible Speech has largely been written off as yet another failed experiment. Yet in contrast to Melville Bell's stymied efforts to promote Visible Speech beyond his immediate family or a small, insular community of teachers of the deaf, in Japan it lay behind Isawa's inauguration of colonial education in Taiwan (1895–1897) and his ideological stance toward national language and its imperial mission. Although it remained largely at the level of methodology, Visible Speech persists in Isawa's varied writings as an auxiliary script to aid in the imperial extension of Japan's "national language" (*kokugo*) into East Asia.

It is worth noting alongside Visible Speech Isawa's other main legacy to Japanese education, namely, the introduction of musical notation. From 1875 to 1877 the Ministry of Education sent Isawa to the Bridgewater Normal School in Massachusetts (now known as Bridgewater State College). It was then one of the leading institutions for Pestalozzianism[11] and state-sponsored national education to the United States. Isawa was already a rising figure in Meiji national education, having served as principal of the

---

11. Swiss education reformer Johann Pestalozzi (1746–1827)'s approach to childhood development applied the theory of *Anschauung* which utilized students' sense-impressions toward forming clear concepts. Eschewing rote memorization, Pestalozzi emphasized teacher-guided object lessons grounded in direct experience. His approaches were widely adopted in the United States in the nineteenth century.

normal schools in Aichi in the early 1870s. At Bridgewater he studied music with Luther Whiting Mason, with whom he established new music training institutes; he also introduced patriotic school songs (*shōka*) and the principle of moral education through music upon his return to Japan.[12]

Isawa spent an additional year at Harvard University before returning to Japan to assume the post of vice principal of the Tokyo Normal School in 1878. He went on to occupy a number of leadership positions in the educational bureaucracy in the 1880s and 1890s. He took over as principal of the Tokyo Normal School in 1879, which he left to become the principal of the Tokyo School for the Deaf in 1886. He brought Mason to teach music on contract at Tokyo University from 1880 to 1882. In 1885, Mori Arinori put Isawa in charge of the Ministry of Education's highly influential textbook bureau, where he replaced the more conservative Nishimura Shigeki.[13] He once again assumed the post of principal at the Tokyo Normal School in 1899 after changing course from his work in Taiwan.

Isawa first became acquainted with Visible Speech at an exhibit about helping the deaf learn to speak at the 1876 Centennial Exposition in Philadelphia. Initially motivated by the desire to improve his own halting English pronunciation, he sought out Alexander Graham Bell in Boston to teach the system to him. Isawa learned directly from Bell, the first and only living "native speaker" of Visible Speech, by engaging in a language exchange of Visible Speech for Japanese. When Bell invented the telephone the following year, he invited Isawa and his colleague Kaneko Kentarō, then a student at Harvard University,[14] to test it out in their

---

12.   Until 1882, Mason (1828–1896) taught in Japan, where he sought to put into practice Pestalozzian educational principles. For a discussion of Western-style music in Japan, including the adoption of the unified style in school songs from 1900, see Eppstein's *The Beginnings of Western Music in Meiji Era Japan*. Isawa was accompanied in the last five months of his stay at Bridgewater by Takamine Hideo (1854–1910), who became an equally important figure in national art education. Analogous to Isawa and Mason, Takamine established Pestalozzian principles in early Meiji art education. See Nakamura, *Shisen kara mita Nihon kindai*, 10–21.

13.   See Twine, *Language and the Modern State*, 160. Nishimura also wrote an article in opposition to Nishi's Romanization proposal in the second issue of the *Meiroku Journal*.

14.   Best known as one of the drafters of the Meiji Constitution (1889), Kaneko was a pivotal figure in introducing the political applications of shorthand notation. He would also be a strong advocate for nationalistic literature, writing the prefaces for the

native tongue, making Japanese only the second language after English to be spoken over the wires.

On the subject of teaching correct elocution to the deaf and stammerers, Isawa likewise maintained continuity with Bell's methods. They remained in contact at least sporadically over the years, as evidenced by the fact that when Graham Bell visited Japan in 1898 to address the Tokyo School for the Deaf, Isawa translated on his behalf.[15] In any event, a crucial distinction that set Visible Speech apart from earlier systems specifically for the deaf and blind such as sign language or Braille was that it was not intended to break up the imagined community of the nation into smaller enclaves. The emphasis placed on surmounting speech impairment by the Bells and Isawa was consistent with their desire to bring regional speakers, colonials, and other marginal subjects into the national, or imperial, mainstream. It is worth noting that in his *Shiwahō-yō onin shinron* 視話法用音韻新論 (The new treatise on phonetics as an application of Visible Speech, 1903), Isawa included a facsimile of a letter he received from Alexander Melville Bell, dated December 11, 1903, commending his efforts to "popularize the system." Bell continues: "I trust that your name may be permanently associated with the introduction to Japan of this purely phonetic system of writing for your native language" (figure 6.4).

Isawa's connection to Anglophone intellectual currents was by no means restricted to his relationship to Melville or Graham Bell. In 1889 Isawa published *Shinka genron* 進化原論, his translation of Thomas Huxley's lectures on evolution. The preface to the Japanese edition, written in English by Edward S. Morse, then professor of zoology at Tokyo Imperial University, contains a striking passage concerning the evolutionary study of language that presciently anticipates Isawa's own concerns for transforming national and colonial language education and his endorsement of social Darwinism as a guiding principle: "The philologist renders the study of language a science, in showing that the many languages at present existing have been evolved from primitive languages of past times. Just as

first Kōdansha collection of *kōdan* (oral tales) published in the late 1920s and 1930s. Akin to Mori, Kaneko was also a correspondent with Herbert Spencer and advanced political views in line with his understanding of social Darwinism.

15.   Likewise, their correspondences were largely based upon the technical scientific grounds of correcting verbal and audial impairments such as stammering. See Go's "Isawa Shūji to shiwahō," 153.

6.4 Letter from Alexander Melville Bell to Isawa Shūji reprinted in *Shiwahō-yo onin shinron*, 1903. Photograph courtesy of Yale University Library.

the modern ideas regarding the origin of language render absurd the story of the Tower of Babel, so does the study of plants and animals, of rocks and stars, render equally preposterous the biblical story of creation."[16] Of course historical linguistics played a crucial role shaping the theory of

16.   Morse, "Preface to the Japanese Translation," in *Shinka genron*, 3.

evolution and was used to justify questions of fitness in the genocides that swept aside indigenous peoples in the name of empire. Historical linguistics was among the first areas of humanistic learning to be defined as a social science. But even if one might trace linguistic origins as though they were geological strata or the taxonomies of the animal kingdom, unlike geology or zoology, language reform was among the surest means for technocrats to effect social transformation.

It was not long before Isawa put into practice his own ideas about the "survival of the fittest" of national languages in the colonial context. As E. Patricia Tsurumi notes, in early 1895 Isawa submitted his proposals to Rear Admiral Kabayama Sukenori, then acting governor-general for the newly acquired territory of Taiwan. Later that year Kabayama appointed Isawa Director of People's Educational Affairs (*minsei-kyoku gakumu buchō*) in Taiwan's civil administration.[17] His emergency and permanent proposals, and subsequent efforts on the ground, were intended to train Japanese and Taiwanese as Japanese-language teachers, and to build up numbers of Japanese-language-capable Taiwanese clerks and interpreters.[18] According to Shi Gang, Isawa regarded Taiwan as an ideal site for the colonial mission. He saw it as a geographic extension of the Japanese archipelago, which shared a common literacy and was in all respects ethnically and culturally "virtually identical" (*hotondo onaji*) to Japan.[19] The first major territorial acquisition after Hokkaidō and Okinawa, Taiwan in many respects became a model colony for radical language policies. This was in no small way facilitated by the free hand that colonial administrators were given in violation of the separation of powers defined by the Meiji Constitution.[20] The curricula reflected

---

17.   This was not the end of Kabayama's involvement with Isawa or education policies. In 1896, Kabayama left the governor-general's office to become the Home Minister; two years later, he was appointed Minister of Education and was responsible for convening the Kokugo Chōsakai (National Language Inquiry Board), which supported implementation of the unified style and other standardizing reforms.

18.   For a useful overview of the beginnings of colonial administration and education in late nineteenth-century Taiwan, see Tsurumi, *Japanese Colonial Education in Taiwan*, 18.

19.   See Shi's *Shokuminchi shihai to Nihongo*, 38–44.

20.   For a more in-depth treatment of Japanese colonial language policies, see Yasuda's *Kindai Nihon gengoshi saikō*. For a more detailed reading of Isawa's dealings in Taiwan, see Tsurumi, *Japanese Colonial Education in Taiwan*, and Tai, "Kokugo and Colonial Education in Taiwan," 503–40.

Isawa's emphasis on inculcating patriotism, morality, hygiene, and physical fitness through military-style calisthenics—all consistent with the philosophy of social Darwinism he espoused.

Isawa was in principle opposed to racism and sought to inculcate through education the cultural heritage shared by Taiwanese, Chinese, and Japanese people, in particular their "common script and literary traditions" (*dōji dōbun*).[21] Nevertheless, there is no question he was a linguistic nationalist who hierarchized Japanese at the center of the Great Japanese Empire. His wish was for the Taiwanese to be integrated not by forcibly rejecting their language, but by learning Japanese properly and in turn having the Japanese study Taiwanese. His views on language and race were at considerable variance from those of Herbert Spencer, who, in an August 26, 1892, letter to Isawa's colleague Kaneko Kentarō, warned against the commingling of races. Spencer replied to Kaneko's question about intermarriage of Japanese and foreigners in no uncertain terms: "It should be positively forbidden. It is not at root a question of social philosophy. It is at root a question of biology. There is abundant proof, alike furnished by the inter-marriage of human races and by the inter-breeding of animals, that when the varieties mingled diverge beyond a certain slight degree *the result is invariably a bad one.*"[22] Building on his advice in previous correspondence with Kaneko regarding the formation of the Meiji *political* constitution based on the unique characteristics of the Japanese people, Spencer adds here a general comment about biological constitutions. He cautions, "If you mix the constitutions of two widely divergent varieties . . . you get a constitution which is adapted to the mode of life of neither—a constitution which will not work properly."[23] Given Spencer's disapproving descriptions of "the Eurasians in India and the half-breeds in America" as well as his opposition to Chinese immigration to the United States, his extraordinary voicing of these opinions to Kaneko was mitigated only by his request that they be kept in confidence during his own lifetime so as "not to rouse the animosity of my fellow-countrymen."[24]

21. See, for instance, Isawa's "Shiwa ōyō Shinago seionhō setsumei," in Shinano Kyōikukai, eds., *Isawa Shūji senshū,* 754–59.
22. Herbert Spencer, August 26, 1892, reprinted in Duncan, *Life and Letters of Herbert Spencer*, vol. 2, 16. (emphasis in original).
23. Ibid., 16.
24. Ibid., 17.

It is unknown whether Isawa knew of these private views communicated to Kaneko and Itō Hirobumi, for whom he served as an interlocutor.
While we may detect some misgivings in his views on racial hybridity in
statements given well after his tenure as the head of colonial education
in Taiwan, Isawa vigorously oversaw the building of normal schools and
implementation of standardized education designed to make Taiwanese
into loyal Japanese subjects. The year 1896 saw the opening of fourteen
primary schools based on Isawa's model, as well as the Japanese Language School, a language teacher's college. Oguma Eiji calls attention
to a speech given by Isawa in 1897 at the highly prestigious and policy-
influencing Teikoku kyōiku-kai (Imperial Education Society) in Tokyo
in which he advocated for an inclusive, multiracial, and multicultural
Japanese society:

> The Native Learning scholars of old held to the interpretation that the great
> people of Japan are none other than the Yamato race, but I believe this is
> a serious mistake. The reverend benevolence of our imperial household is
> by no means restricted to such a narrow compass. In actuality, it is equally
> as vast as heaven and earth. With equality and impartiality for all [*isshi
> dōjin*], [the Emperor] looks at all peoples in the countries of the world as
> his children, whoever they are; provided they submit [to his rule], they
> belong to this great people.[25]

On the other hand, the formation of a *multilingual* society was a more
complicated affair. As Komagome Takeshi argues, the purpose of Isawa's
visit to Tokyo and the substance of his talk were first and foremost to
promote the creation of a six-year primary school and a four-year secondary public school system in Taiwan. This was an ambitious, but enormously expensive enterprise for the Japanese government to undertake
in the aftermath of the Sino-Japanese War.[26] In 1897, once the lifting of
martial law revealed the high costs of empire, there was an inevitable
backlash in Japan against lavish colonial spending, and Isawa had to re-

---

25. Oguma, *Tanitsu minzoku shinwa no kigen*, 70–71. Translation is my own.
26. Komagome, *Shokuminchi teikoku nihon no bunka tōgō*, 44–45. See also his discussion of Isawa's ideology of colonial rule in relation to other figures in the Taiwanese
military and civilian administration, ibid., 51–74.

sign in protest of the deep budget cuts in education that resulted.[27] Tsurumi rightly asserts the longevity of Isawa's legacy in laying the foundations for Taiwanese colonial education: "Under the next governor-general, Lieutenant General Kodama Gentarō, Isawa's ideas were to be realized in a system of schools, curricula, pupils, teachers, and administrators. When Kodama and Director of Civil Administration (*minsei-chōkan*) Gotō Shimpei, arrived in March 1898, sixteen Japanese-language institutes and thirty-six branch institutes were in operation."[28]

The normal schools, which continued to improve in quality and number in the years to follow, were consistent with Isawa's vision that they retain some familiar aspects of traditional Taiwanese and Chinese education such as Confucian ethics (to which loyalty to the Japanese imperial system was adduced) and proficiency in Chinese (necessary for trade with the mainland). Isawa remain involved as a consultant until his appointment as head of the Tokyo Normal School (*Tokyo kōtō shihan gakkō*) in 1899. He was active in promoting colonial education in Taiwan and mainland Asia until his death from tuberculosis in 1917.[29]

While he did not publish his adaptation of Visible Speech called *Shiwahō* 視話法 until 1901, it lay at the heart of Isawa's ideological and educational philosophy, and provided the conceptual center for the seemingly diffuse goals of correcting dialects, enabling deaf education, implementing physical, moral, and musical education in Japanese schools, and actively promoting language studies in the Japanese empire. Despite the relatively late publication date, he had already made considerable use of its methods in the 1880s and 1890s. Go Komei notes that during his tenure as head of colonial education in Taiwan, Isawa published two articles applying Visible Speech to study the phonetic structure of Taiwanese. Moreover, Isawa steadfastly maintained in his later writings that Visible

---

27. Tsurumi, *Japanese Colonial Education in Taiwan*, 18.

28. Ibid. Gotō Shimpei (1857–1929) went on to become the first director of the South Manchuria Railway (1906), Minister of Communications (1908), Home Minister (1916 and 1923), Foreign Minister (1918), governor of Tokyo (1920–1923), and the first director of NHK (Nihon Hōsō Kyōkai; Japan's first public broadcasting company), among other accomplishments.

29. Isawa Takio, who served as governor-general of Taiwan from September 1924 to July 1926, was Isawa Shūji's younger brother, and credited his brother for providing his vision of colonial education.

Speech had permeated his earlier thought and political activities. He taught its principles to Konishi Nobuhachi while serving as the head of the Tokyo School for the Deaf in 1885, and in 1903 founded the Rakusek-isha Company to disseminate Visible Speech to cure verbal impairments such as stammering and deaf muteness. In speeches and essays such as "Shiwa ōyō Shinago seion-hō setsumei" 視話応用支那語清音法説明 (An explanation of Visible Speech for the correct pronunciation of Chinese, 1905) and "Tōhoku chihō hatsuon kyōseihō ni tsukite no ben" 東北地方発音矯正法に尽きての弁 (An argument for a method of correcting the Tōhoku regional pronunciation, undated) he sought to establish Visible Speech as the means of unifying the heterogeneous linguistic conditions within the Japanese archipelago and its colonies.

Although it is beyond the compass of this book to attempt a linguistic analysis of Bell and Isawa's versions of Visible Speech, a few basic observations are possible. Isawa intended for his system to be consistent with Bell's by making the physiological phonetic markers correspond to the table of fifty sounds (gojū onzu) of kana. Sounds originating in the mouth were classified in separate charts for "normal consonants" (futsū fuon) and "normal vowels" (futsū boon), respectively. In the case of consonants, the chart was further divided into aspirate and non-aspirates; the location of the sound-making apparatus (onki) versus the quality of the voice (koe no seishitsu); and so on. For vowels, Isawa was faithful to Bell in grading the distinctions in tone based on placement of the tongue and shape of the mouth.

One of the more significant changes that can be found in Isawa's version was the fine-print inclusion of Japan's colonies into the order of the Japanese sound chart (figure 6.5).[30] Under the heading "Secondary Unvoiced Consonants" (jiseion), Izawa added a note regarding the pronunciation of Chinese and Taiwanese (Shina-oyobi Taiwan-go gaku). It provides another indication of the hierarchical ordering of colonial and colonized languages. Isawa's preface to Shiwahō offers his longest statement on its ideological and practical uses, which echo Bell. It is a study in contrasts between the idealism of establishing a universal language and the strident social Darwinian justification for imperial expansion. Isawa

---

30.   For convenience sake, I have followed Go Komei's lead in collating the several pages on which these charts appear.

reiterates Bell's hopes for Visible Speech to be adopted as the single, best technique for representing all spoken languages in a single "medium" (*baikai*).[31] The preface is, in fact, printed horizontally in kanji and kana yet glossed with its translation into Visible Speech in smaller letters above it. Visible Speech here is thus cofigured as a legitimate alternative to normative Japanese scripts (figure 6.6). Isawa argues that by representing all languages equitably, Visible Speech may yet become a "worldwide standard for phonetic script" (*bankoku futsū onji*).

From the outset, Isawa also tied Visible Speech to the evolutionary theory of natural selection (*shizen tōta*) and the social Darwinian contest between imperial world powers. In this 1901 preface, he holds that "language is humanity's unique weapon" (*kotoba wa jinrui tokuyū no buki*), which sets us apart from the lower orders of animals in our capacity to speak freely, assist one another, and exchange ideas. It therefore follows that language is responsible for the elevation of human civilization. Just as the ancient Chinese texts show that human beings secured their dominion over the birds and beasts with this seemingly mystical virtue, he continues, the survival of the fittest is also an inevitable consequence in the competition of human societies. Those with the best methods of communication, starting with the rational system of Visible Speech, will achieve superiority. But there is an unmistakable shift in emphasis and tone that brings the reader back to the Japanese imperial context.

Expressing words of appreciation to Melville Bell and Graham Bell, Isawa compares Visible Speech to Japan's military reforms and victories over China, and issues a clarion call for Visible Speech to be used to secure the bright future of the Japanese language and people:

Their invention of this system and its instruction to me is akin to the reform of our military science by Westerners who encouraged us to use superior firearms, such that in the Sino-Japanese War in the twenty-seventh year of Meiji [1895] and this year's Peking Incident the martial bravery of our army shocked the world's great powers. In the same way, in the future world conflict of national languages [*shōrai sekai no kokugo*

31.   See, for instance, in "Shiwa ōyō Shinago seionhō setsumei."

五十音

濁　音

次　清　音

6.5 Sound chart to Isawa Shūji's *Shiwahō* (Visible speech), 1901. Photograph courtesy of Waseda University Library.

no kyōsō jyōri ni], the holy light [*reikō*] of our country's language will shine through this invaluable system.[32]

Written by Isawa in the run-up to Japan's next campaign challenging Russia for control of northeast Asia, this bellicose statement demon-

32. Isawa, *Shiwahō*, v–vi.

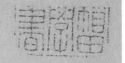

6.6 Preface to Isawa Shūji's *Shiwahō*, 1901. Photograph courtesy of Waseda University Library.

strates how the science of phonetics was applied toward creating a national language that was also from the outset an imperial tongue. In keeping with the militarist framework for the dissemination of Japanese language and citizenship, Shi Gang further points out that Isawa's "Shinbanto jinmin no kyōka hōshin" 新番頭人民の教科方針 (New edition illustrated plan for the people's education, 1897) refers to the Prophet Mohammed wielding a sword in one hand and the Koran in the other.[33] Isawa invoked this iconic image to justify war and education as equally necessary tools in achieving the Japanese imperial destiny.

In the essay "Iwayuru saikin no kokugo mondai ni tsukite" 所謂最近の国語問題に尽きて (Regarding the so-called problem of national language today, 1908) Isawa connected the question of colonial language education to national language reform. He points out problems of orthography and grammar in English, German, and Russian to show that modern Japanese is by no means the only dominant language beset with difficulties. He calls attention to the Autumn 1900 Proclamation on Elementary Schools (*shogakkō-rei*), which included provisions for the use of standardized phonetic reading for pronouncing Japanese syllabary (*jion kana zukai*) and limiting the number of Chinese characters. This was followed by recommendations of the Investigative Committee on National Language (*kokugo chōsakai*) in 1902:

1. To make use of phonographic writing, and investigate the merits of kana and Roman letters;
2. To implement the unified style [*genbun itchi*] as a writing style, and investigate those matters related to it;
3. To investigate the phonetic organization of national language;
4. To investigate dialects and promote standard dialect.[34]

Isawa did not vocally endorse all of these recommendations with equal vigor. He remained largely silent, for instance, on the subject of the unified style, which he used interchangeably with bureaucratic style in classical grammar for his own writings. Despite the impetus for broad reform, Isawa recognized the likelihood of strong resistance to radical spelling reforms such as the abolition of Chinese characters. Setting aside his personal beliefs on the subject, he observes that most Japanese would not stand for changes that decisively broke with Japan's glorious past, such as representing the name of the legendary founder of the Imperial line,

33. Shi, *Shokuminchi shihai to nihongo*, 38.
34. Isawa, "Iwayuru saikin no kokugo mondai ni tsukite," in Shinano Kyōikukai, eds., *Isawa Shūji senshū*, 720–23.

Jimmu Tenno 神武天皇, in an almost sacrilegious vernacularization in kana as ジンムテンノー or in Romanization as "Jimmu-tennō." This recognition of a conservative impulse in the nation did not, however, prevent him from advocating education of the disabled, dialect reform, and colonial education under the aegis of Visible Speech.

Using Taiwan as his model, Isawa actively promoted the value of phonetics and language education for furthering Japan's imperial mission on the continent.

> Taiwan is already a part of Japan. The Taiwanese language is one part of the Japanese language. Yet the Investigative Committee on National Language has made no inquiries about phonetics in Taiwan. Moreover, our armies are deployed in Manchuria and Korea. The opportunity for setting up a foundation for managing Manchuria and Korea will not wait. In its plan for practical education the Ministry of Education has signaled its desire to assemble the personnel necessary for managing Manchuria and Korea. It is with delight that I heard of the training of these persons. We are fast approaching the time when Chinese and Korean must be added as mandatory subjects in our middle schools.[35]

It goes without saying that "national language" here is a multilingual construct in which Japanese supervises and hierarchizes the now-colonial languages of Taiwan, China, and Korea. The purpose of Visible Speech was to maintain the standardized forms of each language so that they could be taught to colonizer and colonized alike.

In his 1902 lecture series "Shiwahō ni tsuite" 視話法について (On Visible Speech)[36] given at the Tokyo School for the Deaf and on two- to three-week-long lecture tours to Iwate and Akita Prefectures, Isawa looked back on his experience of learning Visible Speech from Graham Bell and his difficulties with English pronunciation. He likens his own inauthentic speech to an illegitimate or "mixed race" child: "When I spoke what I thought was English, it was neither pure English nor Japanese. I was speaking something like the mixed-race child [*ai no ko*] of English and

---

35. Ibid., 726–27.

36. Isawa's lectures were recorded in shorthand at those various locations. The surviving version in Shinano Kyōikukai, eds., *Isawa Shūji senshū*, 767–95, was compiled and edited from those transcripts.

Japanese. When both were written down in Visible Speech and shown to me, the difference between them was reflected before me as clearly as if it had been in a mirror." The metaphor of the transparent and reflective surface is telling. By this medium of self-recognition, language use is equated with racial and national belonging. Once the shortcomings of one's linguistic performance was made legible to the speaker, Visible Speech could correct those mistakes through its graphic markers, and reintegrate him or her into the greater national or imperial community. In "Iwayuru saikin no kokugo mondai ni tsukite" Isawa observes that national language is a matter of concern not only for linguists, but has wide-ranging implications for the nation-state as a whole, encompassing issues of governance and the production of academic knowledge. The basis for understanding national language, he insists, is not only through the spoken word, but through the building blocks of letters that are *seen and heard*:

> The national language that has emerged from the midst of today's enlight-ened national citizens [*kaika-seru kokumin*] is not only something that is spoken from the mouth and heard by the ears, but includes an aspect that is written by hand and seen with the eyes. Through a given species of letters this language is written down, and it goes without saying that it is possessed by a necessity that endures beyond the lifespan of the people [*kokumin*]: inherited from past generations and transmitted to our future descendants, it forms the basis of our literature and the original source of our history.[37]

In this statement Isawa brings us full circle in the determination of national language through script reform, akin to Nishi Amane's notion that everything proceeded from the materiality of the letter. Language is defined as visual and inscriptive—that is, as a notation system that permanently records history and provides the continuity for a people to recognize themselves. National script is therefore the external and lasting form of the language by which a nation represents itself to itself. Yet, as Isawa knew only too well, this effect could only be accomplished by re-

37. Isawa, "Iwayuru saikin no kokugo mondai ni tsukite," in Shinano Kyōikukai, eds., *Isawa Shūji senshū*, 716.

mediating the heterogeneity of verbal and visible speech, and by transforming the opacity of "hieroglyphics" into a transparent and standardized phonetics. Under these circumstances, Visible Speech could occupy at best the liminal position of a signpost along the way to the proper relations of national language and script. As we will see in the next chapter, the modern Japanese novel was "discovered" under the same discursive conditions by writing under erasure with shorthand notation.

PART III

"Writing Things Down Just as They Are"

CHAPTER 7

Regime Change

This chapter examines discursive changes in the verbal and visual regimes of the mid-1880s which contributed to the formation of modern Japanese literature and visual culture. In particular it attends to conceptual constellations that represent new writing technologies and the modes of realism they occasioned. As we have already seen, the signifier *utsushi* from the first decade of Meiji came to unitarily encompass the meaning of copy, trace, inscription, and projection in media-technical as well as aesthetic senses. Its semiotic field includes two character compounds that today serve as translations for the Western concepts of realism (*sha-jitsushugi*), sketching (*shasei*), and photography (*shashin*). It has multiple, overlapping relations to phonetic shorthand, from the catchphrase "to write things down just as they are" to its nominalization as both phonography and verbal phonography. Yet this constellation alone did not determine the Meiji episteme. The interpenetrating constellations of oratory, literature, and print that had previously defined the demi-monde of the floating world underwent no less profound transformations in form and sense. Rather than say their relations were decisively sundered, it is better to say that their code-sharing was reorganized into discourse networks of modern media.

The chapter is accordingly divided into three sections. First, it contrasts the modern *utsushi* constellation to Edo woodblock print culture and its basis in calligraphy. It follows with the uncoupling of word and image in consequence of Ernest Fenollosa's *Bijutsu shinsetsu* (The truth of art) which excluded calligraphic literati art (*bunjinga*) and photography from the emerging discipline of fine art.

Second, it attends to the constellation of *hanashi*, or storytelling in the plebian *yose* theaters, and the oratory shifts in the late Tokugawa to early Meiji regimes. It looks to a representative work by Chōrin Hakuen, a popular teller of thieves' tales, who, like Mokuami, got on board with the rhetoric of language reform and government-backed morality campaigns. My concern with Hakuen has less to do with the content of his stories per se, however, than the technical circumstances of their production and publication.

Third, it examines the applications of shorthand in the Meiji political novel *Illustrious Statesmen of Thebes* by Yano Ryūkei. This novel was written in the so-called "translated style" (*yakubuntai*) of predominantly classical grammar, and made no attempt at reform along the lines of the unified style. As such, it was a very different demonstration of the utility of shorthand for quickly recording speech in writing or transforming it into a new kind of literature. Shorthand was unequivocally first applied to a literary text outside the canonical mainstream of realism written in the unified style. It is a valuable corrective to canonical literary history, which ignores a common media history between Yano's political novel and the genealogy that commences with Enchō's *The Peony Lantern*, Shōyō's *Essence of the Novel*, and Futabatei's *Floating Clouds*.

"Utsushi": Between Calligraphy and Photography

Barring a few exceptions, there was little in the way of singular authorship or textual authority in the sense understood by Western aesthetics, and epitomized by Romanticism, until after the Meiji Restoration. In Edo woodblock print culture the multiple hands of production—calligrapher, illustrator, engraver, printer, and so on—made each text a consummately dialogical affair. Accordingly it is not uncommon to see the "signature" of each figure either explicitly or implicitly coded into a given work. For centuries this character, formerly written 寫 or 寫, had been associated with the textual productions of Buddhist monasteries, which, like their medieval European counterparts, encouraged the meritorious act of copying sutras (*shakyō* 写経) and holy books (*shahon* 写本) to preserve and disseminate teachings.[1] This practice occupied its own respected niche in medieval manuscript culture. Edo woodblock print culture, too, enabled skilled calligraphers to earn a living as copyists. The calligraphic hand was an integral part of aesthetic compositions, indivisible from the witticisms, allusions, and elegant turns of phrase devised by the "author." In this sense, calligraphers were not simply hired hands, but artists in their own right. The calligraphers' idiosyncrasies perpetuated a profusion of kana styles in the popular literature of the period: it is their aesthetic choices that render such texts either partly or wholly illegible to contemporary readers who are in fact literate only in modern typography and postwar standardizations of Chinese characters and kana.

Although calligraphy lay at the center of Edo verbal, visual, and print regimes, this was not to last. By the 1890s, the interchangeable and anonymous character of movable type print had displaced woodblock and its basis in calligraphy. By the same token, shorthand and other scripts including Romanization and kana, eroded the utility of the calligraphic hand in the age of phonetic transparency. It would also suffer a substantial loss of prestige at the hand of Meiji art reformers. Photography, oil painting, and even drawing in pencil undermined the primacy of the brush as the principal handheld artistic implement. Nowhere is this more evident than in the case of calligraphic literati art, a consummately amateur practice that cultivates ennobling knowledge of Japanese and Chinese cultural traditions. The combinatory gamesmanship of *haikai*, *surimono*, and other prestige-laden Tokugawa practices are also implicated in this regard. As a reflection of the cultural aspirations of merchants and lower-ranked samurai, literati art perhaps best expressed the interpenetration of verbal and visual regimes: word and image, held together by the same implement, the brush, which etymologically appears in both—書畫: *sho* 書 meaning "to write," and *ga* 畫 meaning "to draw or paint" (now represented by the simplified character *ga* 画). Both characters originate from the ideographic radical for the brush 聿.[2]

---

1. Copying sutras was not only an act of piety, but also an act of meditative concentration in which one not only inscribed the words onto paper, but also learned them by heart.

2. For an in-depth discussion of these terms, see Satō's *Modern Japanese Art and the Meiji State*, 185–86. His explanation of the different permutations of terminology for

Foremost amongst literati art's opponents was the young Ernest Fenollosa (1853–1908), appointed as the first Professor of Philosophy at Tokyo University in 1878. Within a few short years Fenollosa would become deeply involved in debates over the definition of fine art and the future of national art education. He would also be appointed along with his protégé Okakura Tenshin to survey and designate the religious objects in Shinto shrines and Buddhist temples throughout Japan as national treasures.

Although he may not have been aware of the recent semantic inversion in the term *shashin*, Fenollosa railed against the inscription of truth before beauty in his lecture "Bijutsu Shinsetsu" 美術真説 (The truth of art) to the Ryūchikai (Dragon Pond Society) at Ueno Park on May 14, 1882. Founded in 1879, the Ryūchikai was an association of government officials, artists, and intellectuals who were actively involved in promoting Japanese art to the West. The neologism *bijutsu* in Fenollosa's title was a translation word for "fine art" coined for the Japanese delegation to the Vienna World Exposition in 1873, where Japan's exhibit planners sought recognition of Japanese works as fine art as opposed to merely decorative art or artisanal crafts. Fenollosa delivered his speech in English, whereupon it was recorded (precisely how is unknown) by Ōmori Ichū, translated, and published six months later in classical Japanese prose, resulting in a dual attribution on the title page: Fenollosa's oratory (*Fenollosa-shi enjutsu*) and Ōmori Ichū's script (*Ōmori Ichū hikki*).[3] It was this same format that would characterize such works as *The Peony Lantern* and other early Meiji books transcribed in shorthand.

Faced with a no less imminent crisis of artistic virtuosity imperiled by mechanical exactitude, Fenollosa advocated a return to Japanese aesthetics to bring about a new era of Japanese pictures, or *nihonga*, purified of certain perceived heresies such as the intermingling of literature, poetry, and painting in the calligraphic literati art of *bunjinga*. Fenollosa refers

---

e 絵 (pictures), *ga* 画 (drawings/paintings), and *zu* 図 (illustrations) is also relevant to this discursive transformation.

3. No English original is known to exist. It has been Fenollosa's peculiar fate to have two of his main theoretical works—*Bijutsu Shinsetsu* and the posthumous *The Chinese Written Character as a Medium for Poetry*, edited by Ezra Pound—filtered by others. No less uncanny is the radical departure in the interpretation of poetry and painting from the first to the second that Fenollosa, or his interlocutors, took with regard to the status of the phoneme, the hieroglyph, and the visual image.

in the lecture to the work of a "certain German critic," often assumed to be Hegel, whom he taught in classes on political economy, philosophy, and logic and aesthetics at Tokyo Imperial University. Yet the work in question is almost certainly Lessing's *Laocoön*, which argues that poetry and painting must each maintain its own proper unity of idea and form, akin to two sovereign nations sharing, but not violating, a common border:

> The connection between painting and poetry may be compared to that of two equitable neighboring powers, who permit not that the one should presume to take unbecoming freedom within the heart of the dominion of the other, yet on their frontiers practice a mutual forbearance, by which both sides render a peaceful compensation for those slight aggressions, which, in haste or from the force of circumstances, they have found themselves compelled to make on one another's privileges.[4]

This argument would prove highly conducive to Fenollosa's desire to discredit the poetic license of literati art while advocating a new genre of modern "Japanese pictures" (that is, *nihonga*), which would be a synthesis of traditional Japanese materials informed by modern Western aesthetics. Needless to say, it also locates the organization of humanistic knowledge in the same ideological discourse as modern nationalism and imperialism.

Fenollosa discounts the popular notion that mimeticism was the purpose of the arts by contrasting the brute mechanical powers of the camera with the graceful aesthetics of Japanese brushwork:

> If it be true, then the photograph [*shashin*] of the dirtiest and meanest thing in nature must be higher art than the brush painting [*bokuga*] of a beautiful and noble thing, because it is more true. . . . To make the *koto* perfectly sound like groaning or an infant's crying, while it would be like nature, would not be music. So to describe this room truly would not be poetry. In the same way if I copy by painting an inartistic thing in nature, that is not art. Thus, even if I copy nature, I must first define the artistic in nature.[5]

4. Lessing, *Laocoön*, 121.
5. Murakata, trans., "Ernest F. Fenollosa, 'Lecture,'" 57. I have modified her translation somewhat to emphasize the original vocabulary and correct some minor omissions and mistranslations.

This division of truth and beauty was articulated against the indexical relation of photography. Paradoxically, Fenollosa was not opposed to the primacy of the brush in painting; rather, he objected to the violation of disciplinary borders when the brush was also used for poetic composition.

While he was unsuccessful in dislodging photographic realism from Meiji art education or other forms of cultural production, including literature, his attacks against literati art were strongly reinforced by such works as Koyama Shōtarō's "Sho wa bijutsu narazu" 書は美術ならず (Calligraphy is not art, 1882), published within weeks of Fenollosa's lecture. Despite the fact that Koyama was at the opposite end of the ideological spectrum from Fenollosa, he, too, aimed to sever the pair bond of poetry and painting, whose calligraphic cofiguration was now seen to violate Western precepts of good form. This was the beginning of the end for calligraphy's role in national art education. As Satō Dōshin explains,

> The dark days for calligraphy were destined to continue. Calligraphy was removed from the principal ideas defining art that were formed under the Ministry of Education's educational policies. Fundamentally, the ministry's ideas about art were close to a Western model: painting and sculpture ranked the highest, craft ranked lower, and calligraphy was eliminated altogether. A debate took place in Meiji 15 (1882) between yōga [Western-style] painter Koyama Shōtarō, who advanced the view that "calligraphy is not fine art" and Okakura Tenshin, who disagreed with Koyama. Despite Okakura's valiant efforts to restore calligraphy's position, a calligraphy department was never established at either the Tokyo School of Fine Arts or the Tokyo University of the Arts. . . . In government-sponsored art exhibitions, which began in Meiji 40 (1907) calligraphy was never an exhibition category. It was not until the Japan Art Exhibition (Nihon Bijutsu Tenrankai, or Nitten), beginning after World War II, that calligraphy was accepted for exhibition.[6]

Of course, there is a hidden irony in the forced separation of painting and poetry that had been technically and aesthetically held together by calligraphy, which mocks Fenollosa's aesthetics. Calligraphy, which derives from the Greek for "beautiful writing," would be replaced by the manual form of inscription dubbed "verbal photography." Its goal was no

---

6. Satō, *Modern Japanese Art and the Meiji State*, 194–95.

longer to gesture toward inner beauty or essential meaning, but merely to capture, as best its technical limits permitted, traces of the real.

A similar transformation in the chain of signifiers for the notion of the real or true, *jitsu* 実, came to dominate the concerns of Japanese Naturalism from around 1900 into the 1910s, and continued via the I-novel well into the postwar period: realism (*shajitsu*), truth (*shinjitsu*), fact (*jijitsu*), and perhaps most importantly given the tortuous attention to the objective versus subjective perspective of the author, sincerity (*sei-jitsu*). In the long run, however, they, too, were divested of any connection to shorthand, or even the literary sketching movement begun by Masa-oka Shiki and the authors now identified as the mainstream of Meiji-era novelists including Natsume Sōseki, Kunikida Doppo, Shimazaki Tōson, and Tayama Katai. I will take up this issue further in chapter 9.

"Hanashi": Constellations of Speech

In the Edo period, massive discursive changes focused around the het-eroglossia of everyday speech, customs, and manners (*fūzoku*) had already taken place. Despite the often bewildering diversity of spoken dialects and written styles, the political and class-based stratification of Edo-period society kept much of "popular" oral-literary production concentrated in the old downtown area (*shitamachi*) of merchant society and its licensed quarters. Benito Ortolani points out that in the plebeian theaters of ka-buki, bunraku, and *yose* of Edo, Kyoto, and Osaka,

> There occurred a remarkable shift of emphasis with the kabuki theater from mere presentations of singing and dancing to mimicry and the dramatiza-tion of certain types, such as townspeople of various occupations, country bumpkins, and fools. Subsequently, narrators also followed this new trend, and gestures (*shikata*) became a fad. Nakagawa Kiun, for instance, insists in the preface to *Shikata banashi* that in gestures a clear distinction can be made between samurai, farmer, artisan and merchant (*shi-nō-kō-sho*), or between laymen and priests, men and women, old and young.[7]

---

7. Ortolani, *Japanese Theatre from Shamanistic Ritual to Contemporary Pluralism*, 233.

Simply put, the Japanese of the Tokugawa period viewed themselves according to what Bakhtin calls "social birth" rather than ties to a shared identity by blood, soil, or mother tongue. Where they found mirth in the foibles of others different from themselves, they did not find a meaningful sense of horizontal fraternity, as per the imagined community of the modern nation, or any sense of urgency to define an underlying essence of what it meant to be Japanese.

There was already ample precedent for colloquial transcription and the effort to capture oratory using phonetic and/or symbolic markers before the advent of phonetic shorthand. At the turn of the nineteenth century, Hirata Atsutane recorded his sermons using the colloquial *de gozaru* copula by *kōshaku* transcription.[8] Although precise details about this technique are still somewhat murky, Hirata was already in the third or fourth generation of scholars and entertainers to make use of it. Ronald Dore cites critiques by Ogyū Sorai and Motoori Norinaga on the pedagogical circuit of lectures, note-taking, and comprehension in Edo period intellectual circles to the effect that excessive literalism and attempts at accuracy turn "[students] into passive automatons."[9] Sorai acerbically commented upon the indecorousness of students literally hanging on their teachers' every word: "They sit at the teacher's feet and take down every word of his lecture. Word for word, from beginning to end, not a syllable out of place. The worst will even mark a pause where the teacher stopped to clear his throat. They study his intonations and imitate his gestures."[10] These extremes caricatured by Sorai were nevertheless outside the mainstream. Unquestionably there were methods of fast writing that shared basic similarities with late nineteenth-century shorthand. For instance, the brush-wielding scribe takes fewer and faster strokes to represent simplified characters and jots down in symbols idiosyncrasies of speech, gesture, and mannerism. However, as we may glean from Sorai's remarks, these notation techniques were in no sense pedagogically valued or seen as a technology capable of surmounting the stillbirth of Japanese language and ethnos.

Yose theaters and urban street corners also rang with a cacophony of new oratorical performances. In late Edo, the professional narrators

8. See Twine, *Language and the Modern State*, 75.
9. Dore, *Education in Tokugawa Japan*, 140.
10. Ibid., 140.

known as *hanashika*, who were the predecessors to *rakugo* and *kōdan* performers, were free to embellish and improvise upon the work of others. Regardless of how well audiences might know a story such as the epic *Taiheiki*, for instance, the audience's horizon of expectations was contingent upon local storytelling traditions, not inflexible adherence to a canonical text. Arguably one of the most intensive, albeit unsystematic, attempts at media capture of Edo speech in its diversity began with Shikitei Samba's *Ukiyoburo* 浮世風呂 (Bathhouse of the floating world, 1811). Kamei Hideo notes that Samba used new diacritical markers to indicate voiced and unvoiced consonants in the desire to show often minute, but phonetically critical, differences in speech pattern by social class or by regions within and outside of the city.[11] Barbara Cross points out that Samba's use of diacritical marks was not unlike those of puppet play scripts, which indicated how performances were to be chanted or intoned. Furthermore, she contends that he used spatial and graphic markers for the speed and timing of utterances as well as their phonetic qualities.[12] In other words, Samba sought to replicate something of the live performative techniques at play in the theater on the surface of the verbal-visual text in his illustrated prose fiction.

In the first two decades of Meiji, there emerged another constellation whose semantic plurality accrued around oratory: performance (*enzetsu*), public speaking (*kōen*), public address (*kōen*), lecturing (*kōgi*), and so on.[13] From the 1870s onward, shorthand was a constant supplement to oratorical events, whether for political or entertainment purposes, and oftentimes both. The splitting of poetry and painting, and the shift in *shashin* from grasping an essential nature to photography finds an analogous epistemic rupture in the transcription of oratory, which brought about the reduction of *hanashi* 咄／噺 (storytelling) to *hanashi* 話 (literally, "speech") in a tightly scripted sense. Only by rendering the highly skilled speech of *yose* entertainers in a new style of writing could the

11. Kamei, *Meiji bungaku-shi*, 73.

12. Cross, "Representing Performance in Japanese Fiction," 1–20.

13. In *Nihongo no kindai*, Komori describes oratory as "a new kind of media" (*nyū media toshite no enzetsu*), 31–67. It is a misleading characterization that conflates the enunciatory and transcriptive processes, and equates this unity with the "discursive products" that appeared in newspapers, journals, and other publications.

construct of primary orality then emerge from the common speech of the folk and the nation.

The multitude of hands in Edo print culture is met with a multitude of voices in the verbal arts. Nomura Masaaki reminds us of the three dominant oratory modes in the theaters of the floating world: reading/recitation (*yomu*) for *jōruri* and *Naniwa-bushi (rōkyoku)*; narrating (*kataru*) for *kōdan/kōshaku* storytelling; and speaking (*hanasu*) for *rakugo, mandan,* and *manzai*.[14] The differences that obtain among them—for instance, the difference between originating from a literary source text such as we see in the puppet theater versus the spontaneous improvisation of *rakugo*—are flattened or erased when they are made into literary texts.

Working from similar premises, in particular the idea of *rakugo* as a consummately improvisational performance, Kawada Junzō offers a deconstructive interpretation of their respective enunciatory registers.

> We ought to contrast the enunciatory act of *hanasu* 話す (to speak) with *kataru* かたる (to narrate, recount). *Kataru* comes from the same root as *katadoru* かたどる (to mold or shape). Whether something is prelinguistic or expressed in the instance of language, *kataru* traces over (*nazori* なぞり) things that already exist, copying what is given. National language scholars may not support my view, but I regard *hanasu* はなす, which is of more historically recent coinage, as coming from the same root as *hanasu* 放す; that which has no pattern or model to guide it, and therefore bears the meaning of a free utterance. In national script there is the neologism *hanashi* 噺, which is given different shades of meaning with the Chinese characters 咄, as well as those for 話 (speak), 談 (converse), and 譚 (recite, narrate).[15]

Although he cannot prove that an etymological connection exists in the homophonous verbs of *hanasu*, Kawada does an excellent job of teasing out these nuances in *rakugo*'s extemporaneity, or what we might call without too much hyperbole "working without a script." He hits upon a fundamental element integral to the enunciatory framework of *rakugo*, namely, that the performer creates multiple characters and narrative

14. Nomura, *Rakugo no gengogaku*, 13.
15. Kawada, "'Hanashi' ga moji ni naru toki," 2. The characters *hanashi* 噺 and 咄 are neologisms created in the Edo period, not historically recognized Chinese characters.

positions differentiated through subtle shifts in tone, gesture, or facial expression. These effects cannot be sustained, however, when the performance is recorded. The unique variations and improvisations that always occur on stage disappear when bound into a single, authoritative text.

Several problems beset the notion of the shorthand reporter who is physically present at the scene of enunciation and is supposed to "write things down just as they are." First, the reporters worked in teams of two to four and hence produced a dialogic record of the enunciation. Second, it was inevitable that they would confer before or after the fact with the entertainers and politicians whose speech they rendered. This was followed by further revisions by the newspaper or book editor. As these circumstances demonstrate, claims for an immediate, transparent transcription were at best an artificial construct.[16]

Kono Kensuke, who was the first contemporary Japanese scholar to critically consider the implications of Japanese shorthand as a verbal photography, remarks on the fact that shorthand reporters and their editors often fixed mistakes and meanderings in speech to produce a polished text: "Ultimately, [these records] gave rise to the illusion of being a person's actual enunciations which erased [shorthand's] independence as a verbal medium."[17] The shorthand transcriptions and the unified style that followed were never truly faithful or accurate recordings in the manner of the noise captured by subsequent writing machines such as the phonograph. Rather, they were a palimpsest with shorthand as a form of writing under erasure whose partial physical traces exist only on the periphery of canonical texts such as the postscript to *Keikoku Bidan* and the prefaces to *The Peony Lantern*.

16.   Kamei Hideo, for instance, critiques the non-simultaneity, or lack of accord, between shorthand reporters. Drawing upon Wakabayashi Kanzō and Sakai Shōzō's famous transcription of the *Peony Lantern*, he points to such minor discrepancies as the choice between rendering a particular word in kana or kanji (*Meiji bungaku-shi*, 4–6). For Kamei this proves shorthand transcribed materials were collated, interpreted, edited and otherwise textually composed rather than raw data truly "written down as is" (*ari no mama ni utsushitotta*).

17.   Kono, *Shomotsu no kindai*, 154. Li Takanori disputes the validity of Kono's statement about the independence of shorthand as a verbal medium, turning to an example from 1885 quoted in Tanikawa Keiichi's "Koe no yukue" (Whereabouts of the voice) that demonstrates among other issues, the reporters' conflicting accounts of a given speech (*Hyōshō kūkan no kindai*, 126–34).

As has been widely acknowledged in the study of the unified style since the pioneering work of Karatani and Maeda in the late 1970s, the literary notion of "writing things down just as they are" produced new texts and reading strategies ideologically invested in the logic of transparency and immediacy. This process intensified as the unified style that emerged from shorthand transcription was all but erased from the historical record. Kono makes a further critical observation about the indexical relationship of shorthand to realism: "It is highly suggestive that shorthand was known by the metaphor 'the photographic method of words.' . . . Even if we accept that shorthand was the occasion that gave rise to the unified style, the translation process was thrust outside of conscious intent. Once it was grasped as a technology for containing actual enunciations just as they are, this resulted in a mistaken conception of realism."[18]

It is difficult to parse from this terse passage whether Kono is criticizing Meiji writers for investing in this concept of "transcriptive realism" or making a more fundamental claim about the nature of realism—namely, that it never exists outside some ideological and material referent. Either way, the problem of shorthand would be displaced onto other literary categories, the "mistaken conception" passed on as a nonmaterial origin.

The basic characteristics of transcribed works are amply represented in Chōrin Hakuen's *Nasake no sekai mawari dōrō* 情の世界回転燈籠 (Revolving lantern of the sentimental world, 1886), a little-known *kōdan* text transcribed by Shitō Kenkichi (figure 7.1).[19] Perhaps the most famous *kōdan* narrator of his day, Hakuen was a close associate of Enchō, Mokuami, and other performers from late Edo to early Meiji. His works were similarly published with great fanfare. *Ansei sangumi sakazuki* 安政三組盃 (Ansei three exchanges of cups, 1886), for example, was transcribed by Wakabayashi Kanzō and published by the *Yūbin Hōchi Shimbun*. Other works were taken down by leading figures such as Maruyama Heijirō, who coined the phrase "verbal photography."[20]

18.   Kono, *Shomotsu no kindai*, 154.
19.   Now housed in the Maeda Ai Bunko at Cornell University, the book was formerly in the private collection of Maeda Ai.
20.   Other works include *Yukinoyo hanashi: hokuetsu bijin* 雪夜情誌：北越美人 transcribed by Maruyama (date unknown); *Tokugawa Genji ume no kaori* 徳川源氏梅の薫

Known as "Robber Hakuen" in his earlier years for his penchant for stories in the *shiranami* tradition akin to Mokuami's "The Thieves," Hakuen became known later in his career as "Newspaper Hakuen" for his embrace of the goals of Enlightened Civilization. The transcription by shorthand was duly embraced by Hakuen as an extension of the civilizing mission. Consistent with the pattern set by language reform manuals and the first edition of *The Peony Lantern*, Hakuen's preface is devoted to exposing the shortcomings of contemporary Japanese and how they can be overcome through shorthand. He laments that "writing and speech are not one" (*bunshō to kotoba to wa dōitsu narazu*), as they are in the West, and that this condition prevents Japan from achieving the unity of national culture (*ikkoku bunka*).[21] Nevertheless, he insists, "this technique of recording [which] produces a verbal photography" (*kono hikkijutsu ni yorite kotoba no shashin ni nari*) brings Japan ever closer to the goal of "putting in order these spoken leaves of truth, just as they are, without writing anything further" (*sara ni fumi kakazu sono mama koto no ha no makoto wo utsushite kaku wa shitatame*).[22]

In addition to the preface and cover, an illustration depicting Hakuen addressing his audience (figure 7.2) reinforces the changing practices in the *kōdan* theater. Instead of sitting on a cushion on the floor as was customary, Hakuen sits on a chair before a Western table with only a glass of water and traditional fan before him. Several of the men in the audience wear European-style bowler hats. Absent from view, however, are the shorthand reporters and illustrator—the multiple hands who would enable the impression of a single author and text to emerge.

The language of the text is largely consistent with the more commonly studied *rakugo* transcriptions,[23] save the narratological feature that, akin to a play script, notes in parenthesis the first Chinese character from a character's name to designate the speaker. In the next chapter, I will examine how the shorthand transcription of Enchō's *The Peony Lantern* led

transcribed by Sakai Shōzō (1891); *Hagi no tsuyu yamashiro nikki: gōshō yamashiroya wasuke shiden* 萩の露山城日記: 豪商山城屋和助氏伝 transcribed by Ishihara Meirin (1891); and *Misao kurabe sannin musume* 操競三人娘, transcribed by Katō Yoshitarō (1900).

21. Hakuen, *Nasake sekai mawari dōrō*, ii.

22. Ibid., ii–iii.

23. The style of the narration and dialogue is colloquial with sentences ending in the *de arimasu* copula.

7.1 Cover art of Chōrin Hakuen's *Nasake sekai mawari dōrō* (Revolving lantern of the sentimental world), 1886. Photograph courtesy of Cornell University Library.

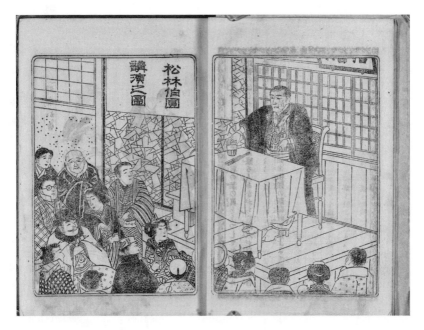

7.2 Illustration from *Nasake sekai mawari dōrō*. Photograph courtesy of Cornell University Library.

the way toward the nascent genre of the realist novel written in the unified style, but first we must turn to Yano's *Illustrious Statesmen* and its use of shorthand as a form of language reform consistent with the nation-building ideals of a political novel set in the golden age of Greek democracy.[24]

True History, Duly Noted: Yano's Political Novel "Illustrious Statesmen of Thebes"

The Meiji political novel was a short-lived genre that reached its apex in the 1880s just prior to the emergence of literary realism. Benjamin

24. *Nasake no sekai* contains an advertisement for a political novel adaptation of the life of King Charles XI. The title is obviously patterned after Yano's *Illustrious Statesmen* (*Keikoku bidan*), entitled *Kaiten iseki Fukoku bidan* 回天偉蹟仏国美談, with the adaptation attributed to Ninten shujin 任天主人閲 and translation by Awaya Kanichi 粟屋関一訳.

Disraeli's *Coningsby* (1844; translated into Japanese in 1884) and the works of Victor Hugo were among the first adaptations and translations of the Western novel in Japan that contributed to fiction's positive reevaluation away from the Neo-Confucianist contempt for gesaku. Christopher Hill has argued persuasively about the relationship between national sovereignty and history that frequently obtains in the political novel as a comparison between states that went beyond the Japanese polity:

> Meiji political novels were not limited to negotiating social change within one national territory, however. They also explored the relationship between a unitary Japanese history (which they were also writing), and the histories of other nations. A prominent example is Tōkai Sanshi's *Kajin no kigū* (Chance encounters with beautiful women, 1885–97), which recounts, among many events, the history of the Carlist rebellion in Spain, resistance to Russian and Turkish campaigns in the Caucasus, the colonization of Madagascar, and the fall of Urabi Pasha in Egypt. The novelists undoubtedly gestured toward other histories to inspire readers with stirring examples like the fall of the Bastille, but in the process they also suggested that the histories of other countries could be abstracted as models for the history of the Japanese nation.[25]

Illustrious Statesmen was no different in this regard, as it expressed the desire to bring about a transformation in Japanese national subjectivity and the governing institutions of state by making recourse to ancient Greece. The novel's commercial popularity enabled Yano to privately fund a trip to Europe to study different political systems and to participate more vigorously in the debates over "the question of national language and script." His research and writing of *Nihon buntai moji shinron* was a direct product of the novel's success.

Illustrious Statesmen, which chronologically and discursively precedes the conceptualization of transcriptive realism in the unified style that is indelibly associated with Enchō, Shōyō, and Futabatei, was an amalgamation of new genres, styles, and scripts. It was based on Yano's translation (*honyaku*) and adaptation (*honan*) of material from a half-dozen English sources on Greek history. Translation and adaptation were closely related

---

25. Hill, "How to Write a Second Restoration," 338.

strategies in early Meiji for disseminating Western concepts and categories of knowledge, especially in literature. In the preface to volume one he expresses his dissatisfaction with the lack of detailed studies of Thebes in English. Although he had originally intended to translate an existing text, when a suitable text did not present itself, he was determined to produce his own. He explains that his purpose was to write an "official history" (*seishi*), with only slight embellishments to entertain and edify the audience: "I added human emotion and humor to lend interest as a novel" (*kore wo hojutsu shi ninjō kokkei wo kuwaete shōsetsu-tai to nasu ni itareri*).[26] But there is another crucial aspect of its composition that set it apart from other political novels: it was the first literary work in the Meiji era to be transcribed in phonetic shorthand. Yano decided to enlist the aid of shorthand reporters to write the novel after an injury to his right arm occurred during his political activities helping to establish Ōkuma Shigenobu's Constitutional Reform Party (Rikken Kaishintō) and the People's Rights Movement (Jiyū Minken Undō), thereby incapacitating his ability to write effectively. He called on *Yūbin Hōchi Shimbun* shorthand reporter Satō Kuratarō to transcribe volume one (1883) and Wakabayashi Kanzō for volume two (1884).

Yano set his sights on recounting the origins of the West in ancient Greece, tracing the rise of Thebes from a tributary of Sparta to conquering its former oppressor and eventually unifying the Hellenic world. Despite the exotic locale, the text was very much consistent with Edo cultural productions set in earlier historical places and times, as per the *sekai* concept in kabuki, which was used to evade censorship and deliver veiled political critiques. Yano, however, sought the intellectual high ground by making his novel historically as well as politically serious. His focus was not simply on reveling in the glories of the Grecian past, but looking forward to the future of the modern nation-state. His novel can thus be read as an allegory of Japan overcoming the unequal treaties imposed by the West, or, more pertinently to his immediate political ambitions, the defeat of the Meiji oligarchs by democracy.

26. Ochi Haruo, ed. *Yano Ryūkei-shū*, vol. 1, 4. See also Maeda Ai, ed., *Meiji seiji shōsetsu-shū* 明治政治小説集 (Anthology of Meiji political novels), the second volume in Kadokawa Shoten's *Nihon kindai bungaku taikei* 日本近代文学体系 (Compendium of Japanese modern literature, 1974), 165–66.

The full Japanese title, *Sēbu Meishi Keikoku Bidan* 斎武名士経国美談, is often rendered by contemporary scholars into English as "Illustrious Statesmen of Thebes." There was, however, already an English translation provided on the original cover by Yano as "Young Politicians of Thebes." Possibly Yano chose a less florid rendition to better communicate its political content to Meiji youth. It was, to be sure, a significant departure from the semi-erotic tones implied by *bidan*, "a beautiful story." Nevertheless, Yano's own translation curiously omits the main part of the title, *keikoku bidan. Keikoku* comes from the *Keikokushū* 経国集 (827 CE), the third oldest anthology of Chinese poetry in Japan. The preface to the anthology contains the phrase *"monjō keikoku no daigyō,"* or "writing is the great enterprise for conducting the affairs of state," which expresses the power of words to regulate political affairs and maintain social and cosmological order. Consequently, even as Yano based his political novel on classical Greece, he also alluded to the Sino-Japanese classics to affirm the value of writing in the service of the state.

John Mertz astutely points out that the narrative is rife with Greek concepts of the public (presented in the novel as *ōyake*), the assembly, and above all, political speaking or oratory.[27] He calls attention to the novel's scenes of the lecture hall, the great assembly hall, and what he calls "the voice of the crowd," which provides the loci of narrative tension and the diffusion of political authority from a single speaker to a community of listeners, that is to say, the audience of the nation. If I may extrapolate a bit further, herein lies the unification of state power with the people. It is a relation expressed by Yano not according to a notion of transcriptive realism or the illusion of a vernacular narrative voice, but through classical eloquence and moral persuasion.

In the preface to volume one, Yano briefly sums up the collaborative process in which he dictated to Satō.[28] Due to the large number of homophones obscuring the meaning of the text, he recounts that Satō would frequently visit him for clarification during his convalescence. Consequently, Yano edited it "by his own hand" (*tezukara*). Neverthe-

---

27. For an overview of the political and narratological dimensions of this text, see Mertz's *Novel Japan*, 208–18 and 230–32. I also strongly recommend his analysis of the marginalization of the political novel from the canonical accounts that begin with Shōyō and Futabatei, 243–67.

28. See Maeda's annotation of *Keikoku bidan* in *Meiji seiji shōsetsu-shū*, 450.

less, he credits Satō and shorthand for enabling the rapid composition of the text. Although only Kamei Hideo has made any mention of shorthand by Satō, Maeda Ai also points out that he wrote a political novel of his own, *Zanfū hiu seiro nikki* (1884), patterned after the adaptation of Bulwer-Lytton's *Ernest Maltravers*, *Karyū shunwa* 花柳春話 (A spring tale of blossoms and willows, 1878–79).[29] Satō published his novel under the literary pseudonym Kikutei Kōsui 菊亭香水, and followed a more conventional approach to the political novel than Yano. More work needs to be done in this area, but I should note that Satō is mentioned in connection with Yano's *Keikoku bidan* in the preface to another similarly named political novel, Katō Masanosuke's translation-compilation (*sanyaku*) *Eikoku meishi kaiten kidan* 英国名士回天綺談 (1885). In this divergence the full contours of the early Meiji novel from the "mainstream" codes of the political novel to Yano's experimentalism to the shorthand transcription of rakugo and kōdan become apparent: a discourse network of transcription as well as a genealogy of texts.

Yano also mentions the artist Kamei Shiichi, whose lithographic illustrations appear in both volumes. While effusive in praise of Kamei's ability to "capture the appearance" (*arisama wo moshitari*) of historical figures and customs of ancient Greece, Yano uses the illustrations more or less in the manner of Western fiction merely to visually reinforce the verbal narrative's depiction of monumental architecture and democratic assemblies. While he eschewed the sort of minute descriptive language Tsubouchi Shōyō (1859–1935) would advocate for in lieu of illustration in *Essence of the Novel*, Yano makes a clear break with the word play and word-image co-figuration of *gesaku*. Lastly, returning to his comments about the text as both a history and a novel, Yano makes a preemptive objection in the preface against labeling his work a "popular historical novel" (*haishi shōsetsu*). From his point of view, this was a disingenuous genre that invented fictitious worlds instead of accurately portraying this one. Simply put, the political novel that Yano had in mind therefore had no precedent in terms of genre or composition. What makes this comment more provocative is the fact that the transcription of Enchō's *The Peony Lantern* was commissioned by the Tokyo Haishi Shōsestu Shuppansha (Tokyo Popular Historical Novel Publishers).

29. See the annotation of *Keikoku bidan* in *Meiji seiji shōsetsu*-shū, 450.

In the preface to the second volume, Yano's interest in script reform is more explicit, echoing the linguistic concerns he addressed in *Nihon buntai moji shinron*. Under the separate heading "On Style" he provides a series of simple observations about literary origins: "Before *Sashiden*, there was no *Sashiden* style. After *Sashiden* came out, then for the first time there was a *Sashiden* style."[30] He repeats this rhetorical gesture with respect to the *Tale of Genji*, the *Shiki* 史記 (in Chinese, Shih Chi; Records of the grand historian, second century BCE), and the *Taiheiki* 太平記 (Chronicle of the great peace, late fourteenth century), remarking that style is not a question of age, but innovation. Before a style may be identified as such, there must first be a unity (*ittai*) to transmit and reproduce. Still, the formation of a literary style does not necessarily beget others, nor is it free from the conditions of its time (*jizoku*).

Yano then describes his own efforts to analyze and contribute to the linguistic reforms since the Imperial Restoration. He expounds upon the four dominant modes of literary writing in Japan in the 1880s: "Japanese" (*wabuntai*), "Classical Chinese" (*kanbuntai*), "vernacular speech" (*zokugo-rigen*), and "direct translation from European languages" (*ōbun-chokuyakutai*) also known simply as the "translation style" (*yakubuntai*).[31] The last is a mixture of the previous three, and as he remarks "from the standpoint of conservative-minded writers of only one style, it must in fact seem to have a monstrously strange [*kikai genyō*] form."[32] By everyday logic, he continues, it makes more sense to use one instrument or apparatus (*kikai*) than to try to combine four at once. In this respect, well in advance of Ōsugi Shigeo's play on the polysemy of Mori Arinori's name *yūrei* (ghost) discussed in chapter 4, Yano anticipates the motif of the ghostly homophone caught in the machinery of writing.

Yano admits that he tried to use all four styles in the second volume, in particular the vernacular style. Unfortunately, the more diligent his efforts, the more they met with laughable results, until he finally gave up and just wrote as came naturally to him. His inability to create a vernacular style notwithstanding, Yano introduced shorthand as a compositional, if not yet explicitly literary, strategy for "writing things down just

30. Yano, *Keikoku bidan*, 2:3. Translation is my own.
31. Ibid., 2:6. Translation is my own.
32. Ibid., 2:8. Translation is my own.

as they are." Much as the preface deals with style and language, the postscript pertains to the utility of scripts. He contrasts Takusari's newly adapted shorthand to the more widespread, workaday uses of the translation style:

> Whether it is used in the court, a social gathering or my request today for precise note-taking, there are many needs for this kind of [shorthand] practice. Yet in many cases, the form of note-taking is that of "translated writing in Chinese" (*kambun yakubuntai*), not words written down precisely as they are spoken (*kotoba wo sono mama seimitsu ni hikki suru mono ni arazu*). No matter how carefully one employs the translated Chinese style, it never gives any evidence of speech as it was enunciated *(kesshite hatto seru kotoba no chokushō to nasu ni tarazu)*. This is one of its major shortcomings.[33]

While it would be tempting to enfold his comments into the discourse of the unified style, it is probably more accurate to say that Yano recognized shorthand as a technical accomplishment, not as a literary style in its own right.

This would appear to be borne out by the postscript to the novel. Yano had Wakabayashi write out the first several lines of the text accompanied by kana-only and kanji-kana script (figure 7.3). We must remember that there was no unified Japanese language at this time, and the status of shorthand as a supplement or alternative national script had not yet been foreclosed. Setting aside his injury as a rationale for turning to the prosthesis of shorthand, Yano reveals an interest that preceded Takusari's adaptation of shorthand:

> I thought it a shame that there was nothing in Japan like Western shorthand notation, and I explained to my acquaintances about the need for its development. But some time later I heard about [a group of] people who were experimenting with shorthand. After trying to decide how best to support their efforts, I settled on asking for their assistance in quickly completing this section of the book.[34]

33. Ibid., 2:Postscript, 1. Translation is my own.
34. Ibid., 2:Postscript, 1. Translation is my own.

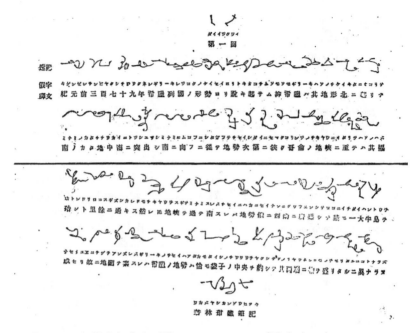

7.3 Postscript to *Keikoku bidan* (Illustrious statesmen of Thebes) in three scripts. Photograph courtesy of National Diet Library.

In the postscript, the three scripts appear side by side for the first time to a larger reading public. Shorthand is made visible as the phonetic, if still indecipherable, ür-script beneath the surface of the text. As we will explore further in the next chapter, it was as writing under erasure that shorthand was divested of its claims to phonetic transparency and transcriptive realism in the formation of the modern Japanese novel.

The term *genbun itchi*, coined by Mozume Takami in 1886, began to gain currency only after these constellations had coalesced into a new discourse of modern Japanese language and literature. At the risk of polemical repetition, I must once again state my objection to imposing the teleology of the unified style back onto the mid-1880s, when it was by no means a universal goal for literature any more than it was for national language and script reform. Instead, I seek to closely examine literary practices and concepts through the medium of shorthand that brokered the possibility of the unified style through its co-figuration with transcriptive realism. What emerges from the disparate fields of judicial and parliamentary reporting, public speaking, *rakugo*, *kōdan*, and so forth is the rapid convergence of media, language, and a discourse of realism that "captures things just as they are." Faith in the commensurability of speech and writing, including the cognate belief that phonetic scripts can capture an actuality rather than an approximation of the scene of writing, was predicated on these ideological and material developments.

In keeping with the comparative focus on the Anglophone world and globalizing developments touched upon earlier, we might note in passing the spectrum from orality in general to particular oratory practices in the late nineteenth century. Lisa Gitelman observes that the profusion of the spoken word oftentimes set the context for vast fields of textual production in the United States.

> Though relatively few Americans had any direct experience with shorthand, literates all had some contact with the underlying matter representing orality. Children labored noisily over spelling books. Readers consumed published oral forms—lectures, sermons, and trial reports—while the more fictive genres of American literature frequently appropriated or invoked the orality of yarns and tall tales. Literary authorship required a good ear, most notably as it was practiced by Mark Twain or John Gregory Dunne, or by regionalists who relied upon the exaggerated verisimilitude of printed dialects.[35]

There was a concurrent shift in the value of classical learning, including the eloquence of oratory, in tension with the figure of the machine. Following Leo Marx's pioneering work in *The Machine in the Garden* on the dueling impulses of the pastoral and industrial in the American national imagination, Eric Cheyfitz maintains that "until the 1850s in the United States the mechanical arts sought to justify themselves in terms of the literary arts. . . . However, as we know quite well today, it will be the literary, or rhetorical, arts that will have to justify themselves in terms of use, or practicality, which the machine will have appropriated to itself

---

35.　Gitelman, *Scripts, Grooves, and Writing Machines*, 52.

from the historical realm of language's potency."[36] Among its many con-
sequences, this shift leads to the disavowal of the mechanical, or nonliterary, origins of literary production. On the one hand, the novel, which
in the eighteenth century was a popular form indistinguishable from the
romance or tale, was privileged by the projects of nationalism and modernization, albeit not always in concert, above poetry as the dominant
form of literature. On the other, the aspects of the novel proper to the
mechanical arts were invariably marginalized by the processes of canonization that defined such works as *jun bungaku*, literally meaning "pure
literature."

36. Cheyfitz, *Poetics of Imperialism*, 31.

CHAPTER 8

The Haunted Origins of Modern
Japanese Literature

The Transcriptive Realism of Sanyūtei Enchō's
"The Peony Lantern"

Modern Japanese literature begins with a ghost story. Transcribed in pho-
netic shorthand and published as a series of pamphlets in 1884, Sanyūtei
Enchō's *rakugo* performances of *The Ghost Story of the Peony Lantern* were
immediately upheld as a model for a new realist novel in vernacular prose.
The genealogies of modern Japanese literature that commence with *Essence
of the Novel* as Japan's first work of modern literary theory and Futabatei
Shimei's *Floating Clouds*[1] as "Japan's first modern novel" have long signaled
The Peony Lantern as a source of fleeting inspiration, but have otherwise
bypassed its composition in shorthand, leaving it to haunt the margins of
the canon as a ghostly remainder. Even in Karatani Kojin's *The Origins
of Modern Japanese Literature* there is almost no mention of these groundbreaking developments in the 1880s, only the fait accompli of new modes
of realism and naturalism, which, from the 1890s onward were written
in the unified style and which produced such epistemic ruptures as the
"discovery" of interiority, landscape, and so on. Although the past two

---

1. *Ukigumo* is often translated as "Floating Cloud," which presumes the singularity
of its protagonist, narrator, and/or author. I have deliberately chosen to render it in the
plural to reflect shifting narrative perspectives, a multitude of different genre conventions, and its verbal-visual dialogic composition.

decades have witnessed new media studies on the formation of modern language and literature, none has yet succeeded in overturning the dominant paradigm of a modern literature divested of its media-historical origins.

While I would once again signal my debt to the groundbreaking research already done in Japan, I offer two contentions that have not yet been articulated by previous scholars with sufficient methodological or archival backing. First, shorthand notation was instrumental in revealing for the first time the possibilities of modern literature and so-called vernacular writing to the Meiji reading public. Known in Japan as both phonography and "verbal photography," shorthand was a manual technique of writing that conceptually embraced existing notions of high-fidelity recording and transmission prior to the advent of mechanical sound-recording technology. As such, it played an indispensable role in bridging, at least at an ideological level, the incommensurability of speech and writing. Second, the publication of Enchō's *The Peony Lantern*, Shōyō's *Essence of the Novel*, and Futabatei's *Floating Clouds* took place amidst sweeping discursive transformations in the interpenetrating oral, visual, and literary regimes of the Tokugawa period. Accordingly, it might be more accurate to say modern Japanese literature begins with the *multiple postings* of a ghost story—a statement more inclusive of transmission across various languages, media, and genres. As we peel back the layers of intertextual and intermediary composition, the media processes that were effaced from canonical accounts of these three texts come into their own as possessed with historical agency.

The multiple layers of literary translation and adaptation that went into the modern *rakugo Ghost Story of the Peony Lantern* prior to its transcription in shorthand are a case in point. The medieval Chinese ghost story *Sentō Shinwa* 剪灯新話 (Tales by candlelight, c. 1378) was first adapted in Asai Ryōi's collection of ghost stories *Otogibōko* 伽婢子 (1666), which in turn provided Enchō with the source text for his theatrical performances. Nonetheless, it was hardly a slavish imitation. In addition to the liberties he took to make it his own, *The Peony Lantern* was narrated in the Fukagawa dialect of Edo-Tokyo, which was closely associated with *gesaku*, plebian theater, and the lively voices of the licensed quarter. Prior to its emergence, via shorthand, as the basis for the unified style in literature and the standardized dialect for nation and empire, this dialect was first and foremost the language of urban sophisticates and the Edo townspeople.

Enchō's performances were a tour de force of verbal and gestural storytelling that disappeared at the end of each night's show. *The Peony Lantern* accrued its material layers one after another through the advent of modern writing technologies. Transcribed by Wakabayashi Kanzō and Sakai Shōzō in shorthand, the installments of the story were edited, typographically composed in mixed kanji-kana script, and published in serialized pamphlets by the same Tokyo Publishers of Historical Tales as Yano's *Illustrious Statesmen*.

Ultimately, my analysis here, too, is less concerned with the substance of Enchō's ghostwritten ghost story, his intentions as the putative author of the text, or in recapitulating the history of Meiji print culture. Rather, it is to address how the shorthand transcription of *rakugo* produced the twin hallmarks of modern Japanese literature: phonetic transparency and mimetic realism. Indeed, the compositional imperative of shorthand practitioners to "write things down just as they are" was co-opted as a rallying cry by the mainstream of modern Japanese literature from the 1890s onward.

It is difficult to overstate the degree to which the various genres of early Meiji—adaptations, translations, transcriptions—were steeped in popular Western literature.[2] Enchō was hardly exceptional for his lack of ideological purity when it came to preserving Japanese aesthetic traditions. He drew freely from whatever sources were likely to be well received by his audience. The source materials for Enchō's many other shows were similarly varied; they were just as likely to be recent English penny dreadfuls as Chinese tale literature.[3] Incongruously enough, one of the earliest works of Western Naturalist fiction to circulate in Japan was Guy de Maupassant's *Un parricide* (1884), adapted by Enchō as *Meijin Chōji* 名人長二 (Chōji the master), not as a shorthand transcription of his *yose* performances, but also as a text written in his own hand. It was first

2. On the relations between transcriptions, translations (*honyaku*), and adaptations (*honan*), see Miller's *Adaptations of Western Literature in Meiji Japan*.

3. See also Miller's analysis of Enchō's adaptation of Charles Reade's novel *Hard Cash* (1863) into *Seiyō Eikoku kōshi Jōji Sumisu no den* (A Western romance: The tale of George Smith, dutiful English son, 1885) in *Adaptations of Western Literature in Meiji Japan*, 85–109.

published by the *Chūō Shimbun* (Central newspaper) in 1898, and re-printed the following year by the Hakubunkan publishing empire.[4]

While *The Peony Lantern* may have been Enchō's first tale to be pub-lished in shorthand, it was not the first to appear in print. A number of his early works were published as "storybooks" (*hanashibon*) from 1864 to 1884 by the early Meiji *gesaku* writer Jōno Denpei, also known by the sobriquet Sansantei Arindo.[5] Enchō thus straddles the discursive bound-ary between late Edo and early Meiji and their fast-changing oral, visual, and literary regimes. While it is not clear precisely how those first stories were written down, it is certain they did not follow the model of verba-tim media capture associated with shorthand.

Of Enchō's "shorthand-transcribed books" (*sokkibon*) dated after 1884, roughly half are attributed to Wakabayashi Kanzō and/or Sakai Shōzō as the principal reporters.[6] The transcripts were published in news-papers and as books and collected works sometimes running upward of a dozen volumes, as was the case with Shunyōdō's publication of Waka-bayashi's transcription of Enchō's *Shiohara Tasuke ichidaiki* 塩原助一代記 (The biography of Shiohara Tasuke, 1885–1886). The monthly shorthand journal *Hyakkaen* 百花園 (One hundred flowers), published from May 1889 to November 1900, featured seven or eight serial installments of *rakugo* or *kōdan* transcribed in shorthand per issue. Some fifty racon-teurs appeared in its pages, although Enchō did not appear to play a prominent part. Phonographic recording in the Edisonian sense came only three years after Enchō's death in 1905 when his former colleague and rival, the Australian-born raconteur Kairakutei Black, was appointed by London Gramophone to assist in recording *yose* shows.[7]

The Peony Lantern was performed in a cycle of forty-six days, with episodes first published as short pamphlets. The tale begins by harkening

4. See Keene's *Dawn to the West*, 224–25.

5. Jōno was the father of the artist Kaburagi Kiyokata, who illustrated many Meiji novels and literary journals. In the 1930s Kaburagi painted a portrait of Enchō, with whom he was well acquainted in his childhood, in the *shin-nihonga* style. On the first Enchō hanashibon, see Itō, ed., *Sanyūtei Enchō to sono jidai*, 59–63.

6. See Itō, ed. *Sanyūtei Enchō to sono jidai*, 61–63. Interestingly, none are attributed to Takusari Kōki, the founder of Japanese shorthand.

7. Ortolani, *The Japanese Theater*, 258. Black's *rakugo* performances were also exten-sively recorded in shorthand in the 1880s and 1890s.

back to the heyday of late Edo culture and establishes a nostalgic distance from contemporary Meiji: "On the 4th May 1743, in the days when Tokyo was still called Yedo, the festival of Prince Shotoku was celebrated at the Shinto temple of Tenjin in Yushima, and the worshippers assembled in great crowds on the occasion."[8] This preamble is uncommon for *rakugo*, in which events are not typically tied down to a specific time or place. Still, it is by now possible to say that the story was no longer structured solely for theatergoers, but for readers of serial fiction. It was one of several narratological concessions that fixed the fluidity of *rakugo* narrative in the manner of a play script or novel. Another is the identification of each speaker on the page by the first Chinese character of his or her name, whereas in *rakugo* an instantaneous shift in tone or mannerism alerts the audience to a corresponding shift in persona.

The multiple hands involved in the composition of the shorthand-transcribed text are likewise represented in a manner consistent with Edo popular literature. In the colophon of *gesaku* works, the author is typically marked by the suffix *shiki* or *jutsu*; the calligrapher with *hikki* or *shirushi*; the illustrator with *ga* and so on.[9] The cover of *The Peony Lantern* is signed "oratory by Sanyūtei Enchō" (Sanyūtei Enchō *enjutsu*) and "transcription by Wakabayashi Kanzō" (Wakabayashi Kanzō *hikki*) (figure 8.1). From the 1890s onward, authorship of texts would be restructured into the same Romantic and post-Gutenberg conceit of origins and originality that hold today. The mid-1880s record of a split between the voice and hand of the author in shorthand transcriptions therefore marks a critical transition between the collective production in manuscript and woodblock print cultures on the one hand, and the modern literary-typographic formation on the other.[10]

Similar to the postscript of Yano's *Illustrious Statesmen* published several months earlier, an advertisement reprinted on the first page of

---

8.   Chamberlain, *A Handbook of Colloquial Japanese*, 447.

9.   This is true regardless of genre—*yomihon, kokkeibon, sharebon,* and so on, all authenticate their means of production in the same way.

10.   Much as we tend to ignore the collective work of film crews to focus upon a single director, it would be absurd to think modern literature does not continue to involve implicit and explicit participation by editors, illustrators, ghostwriters, and others. Nevertheless, ideological consistency requires that these issues be subsumed in favor of the dominant image of a stand-alone author.

8.1 Cover of *Kaidan botan dōrō* (Ghost story of the peony lantern), 1884. Photograph courtesy of National Diet Library.

each serial pamphlet of *The Peony Lantern* included a demonstration of shorthand alongside mixed kanji-kana script by Wakabayashi for the edification of the reading public (figure 8.2). It reads:

> This *Ghost Story of the Peony Lantern* is an adaptation based on a famous Chinese romance that provides a new twist on a ghost story. Not only is it extremely exciting; it is a tale where good is rewarded and evil punished. It is one of the sentimental stories at which Master Enchō excels, and he performed it every time to great applause.
>
> *Kono kaidan botan dōrō wa, yūmei naru Shina no shōsetsu yori honan seshi shinki no kaidan ni shite, sukoburu kō aru nomi ka, kanchō ni hieki aru monogatari nite, Enchō-shi ga tabi ni chōshū no kassai wo hakuseshi, shi ga tokui no ninjōbanashi nari.*[11]

This advertisement offers some valuable insights into popular literature in early Meiji. Wakabayashi describes *The Peony Lantern* as a Chinese novel, tale, and romance, denoting the permeable genre boundaries prior to Shōyō's *Essence of the Novel*. While Wakabayashi unabashedly embraces the fact that the text is an adaptation, he also stresses the "novelty" (*shinki*) of Enchō's performance that sets it apart from previous versions. Still, as per McLuhan's famous adage, the medium is the message. The advertisement was less a demonstration of shorthand in order to sell books than a spectacle in its own right, as if to say, "This is what the transcribed text of a ghost story looks like beneath its skin." This unsettling, spectral vision of the phonetically transparent medium was covered up by subsequent editions of text, and even omitted from most editions of collected works for literary specialists once the shorthand connection to literary realism was disenfranchised by canonical literary studies.

Wakabayashi's preface to *The Peony Lantern*, which has been included in most modern typographic reprintings of the text, offers a more detailed account of his participation in the transcription process. Commissioned to work with Enchō by Tokyo Historical Tale Publishers, Wakabayashi left behind the high-minded rhetoric of Yano's *Illustrious Statesmen*, which he had transcribed only months earlier, to work on a twice-told ghost story

---

11.   Itō, ed., *Sanyūtei Enchō to sono jidai*, 30. Translation is my own.

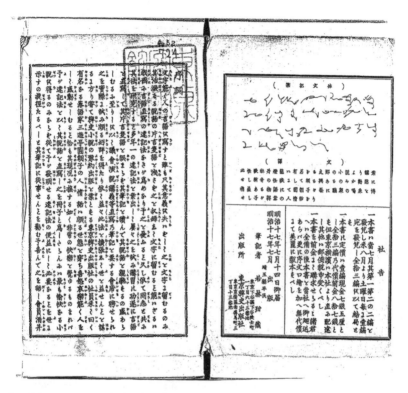

8.2 Advertisement with shorthand script in the first edition of *Kaidan botan dōrō*. Photograph courtesy of National Diet Library.

steeped in human passion. Never before translated into English, the preface also reflects awareness of the technologies of the letter and burgeoning rhetoric of shorthand as phonography and verbal photography. I include the preface here in its entirety.

Although letters transmit well the words people speak, if they lose their meaning, they remain nothing but letters. The fact that I was able to record this lively story without reducing it to gibberish or mere scribbling is due to our nation's methods of shorthand notation for transcribing words. It has been a long time since I felt anxious over this. For many years I researched the methods of shorthand notation with my comrades before devising a system of my own. I undertook many experiments and the result of those trials was the transcription of words without mistakes that would lead to gibberish or nonsense [*hengen-sekigo*]. Reading the script I'd written imparted the feeling of having faithfully listened to the story. I would sit in attendance wherever there was a need for written transcription, whether for the Diet, an oratory performance, a lecture, etc. In fact I received a tremendously positive response to these experiments. This further encouraged me to expand my method and endeavor to show the world its utility.

As an editor from the Tokyo Historical Tale Publishers suggested to me, the famous storyteller Sanyūtei Enchō's sentimental stories admirably express the richness of worldly affairs, touching us with the four emotions of delight, anger, sorrow, and happiness. Striving to bring the incomparable pleasure I experienced as close as possible to the actual performance, I employed shorthand to transcribe this story. What you see before you in this pamphlet is not only a pleasurable novel [*yukai naru shōsetsu*], but an expedient means for me to demonstrate to the world the necessity and utility of the shorthand I invented. In addition, it was a means of applying my skills. I happily accepted [the publisher's offer] and together with my colleague Sakai Shōzō, I went to the *yose* theater where Enchō gave his performances. Sitting in a room offstage, I used shorthand to transcribe Enchō's performance just as it was delivered [*sono mama ni chokusha shi*], without having to correct for any nonsensical remarks. And so it was published, *The Ghost Story of the Peony Lantern*. It was adapted from a renowned Chinese novel to make a thrilling new ghost story. Not only is the story extremely interesting, it is a didactic tale that rewards good and punishes evil in a way that always draws the audience's applause. As Enchō skillfully related this sentimental story, I listened as always while observing the actual situation [i.e., his performance] and wrote without distortion. If he laughed, my words also laughed. If he grew angry, my words also grew angry. Tears were met with tears and delight with delight. The words of the maiden are gentle and polished, while those of the strapping young man are blunt and down-to-earth. Since they were recorded using so-called "verbal photography" [*kotoba no shashinhō*], I believe readers of this pamphlet will experience the same pleasure as if they had faithfully listened to Enchō's sentimental tale in the theater.

It is my hope that this pamphlet will expand knowledge about the utility of my system of shorthand. With shorthand, when the transcribed words here and there fail to produce a good tone, it is unconducive to smooth reading and the story will not seem like a normal novel [*jinjō shōsetsu*]. In other words, we see at such times a type of shorthand that directly transcribes words without capturing their tone, as well as the impoverished grammar in the tales of our country. If I may claim a larger objective here, it is to contribute to our country's future linguistic improvement. I hope those of you who have seen Enchō firsthand will enjoy reading this pamphlet.[12]

In contrast to the rhetoric of "writing things down just as they are" seen in various shorthand manuals and Yano's *Illustrious Statesmen*, transcribed speech no longer appears in a strictly technical capacity of mechanical accuracy, but as an aesthetic category. From the outset, Wakabayashi targets for elimination homophones that might disrupt the circuit from authorial voice to inscriptive hand to transparently readable text. In harmony with the notion of a writing system in which form precedes meaning, he lauds verbal photography's capacity to perfectly capture a *rakugo* performance such that the reader should feel as though he or she were present in the theater. In this powerful call to affective immediacy, Wakabayashi invokes the hallucinatory effects of the transcribed words that seemingly leap off the page and come to life. Shōyō would return to this principle as the touchstone for the realistic depiction of sentiment and for grasping the essence of literary hermeneutics.

Despite his dialogic effort with Sakai Shōzō and, of course, Enchō, Wakabayashi denies personally adapting or interpreting the text. He evades what Lisa Gitelman identifies as the central paradox of shorthand—namely, that "it was enmeshed within a rhetoric of progress that cast the reporter as a technician, contradictorily both skilled and automatic."[13] It is instructive to compare Wakabayashi's approach to *The Peony Lantern* with Shikitei Samba's brief comments at the start of *Bathhouse of the Floating World*. The historical connections between *them* are manifold. Enchō was only one or two generations removed from Sanshōtei Karaku, the founding father of *otoshibanashi*, the forerunner to *rakugo*. Indeed, Samba credits Karaku's incomparable humor with having inspired his text. On a separate tack, in "Yo ga genbun itchi no yurai" 余が言文一致の由来 (The origins of my genbun itchi, 1902), Futabatei

<hr>

12. Wakabayashi, "Preface" to Enchō, *Kaidan botan dōrō*, unnumbered first page. Translation is my own.

13. Gitelman, *Scripts, Grooves, and Writing Machines*, 60.

Shimei relates how he not only relied upon Enchō, but also looked to Shikitei Samba for clues how to represent spoken language.[14] Notwithstanding Samba's self-deflating claim to have "farted forth this little volume,"[15] there is a crucial difference between Wakabayashi and Samba's modes of representing enunciation: where Samba marks his narrative with a lament for lost content and the failures of memory to restore to presence the stories he heard, Wakabayashi insists upon rigorous, word-for-word transcription. Samba explains:

> One evening in Utagawa Toyokuni's lodgings, we listened to Sanshōtei Karaku telling [*otoshibanashi*] stories, and, as always, his gifted tongue went right to the heart of human emotions. There is surely no one as funny as he; how hard it is to be even a tenth as effective on paper! Beside me that evening, laughing as hard as I was, sat a publisher. Greedy as ever, he suddenly asked me if I would put something together based on these stories of the public bath.[16]

Where Samba could claim to recover only something like 10 percent from what was lost in transmission, Wakabayashi insists that shorthand captures the actuality of what was said rather than essence of its meaning.

Tsubouchi Shōyō's preface to *The Peony Lantern*, written under the pseudonym Harunoya Oboro, not only pays tribute to shorthand and the use of the vernacular in the composition of the text, but also lays a far greater emphasis on the literary qualities of Enchō's stories of sentiment in contrast with the fictional masterpieces of Tamenaga Shunsui and Samba.

> "Those who are able to express sentiment (*nasake*, glossed *kokoro*) almost exactly as it is thought instinctively master the art of writing by naturally fulfilling the rules of rhetoric," wrote Spencer in his later years. Truer words were never spoken! Recently, the ghost story teller Sanyūtei [Enchō] gave an oral performance of the story called *The Peony Lantern* that was transcribed just as it was through a method called shorthand notation (*sokki to*

---

14. Kōno and Nakamura, eds., *Futabatei Shimei zenshū*, 9:148–49.
15. Leutner, *Shikitei Sanba and the Comic Tradition in Edo Fiction*, 139.
16. Ibid., 141.

iu hō ni mochiite sono mama utsushitori). If you glance at the written page, you will see it only makes use of provincial and colloquial language. Since it is not given to flowery expressions, the effect becomes increasingly more dynamic with each phrase and sentence. It is as though we come face to face with the man Hagiwara, and one has the feeling of actually meeting the maiden Otsuyu. Reading of Aikawa's impetuousness and chivalrous conduct alternatively makes one laugh or feel moved in spite of oneself, but it comes across so strikingly true that one cannot think of it as a mere story. Is this not an extraordinary effect of such base writing? Even if it [the dynamic style] truly comes about from the ingeniousness of the writing, Enchō has never been part of the literary establishment, and he has never studied the craft of the written word. Yet as we can see from the way he composes lines from each word and breath, he gives Tamenaga [Shunsui] a run for his money, and outfoxes Shikitei [Samba] in his mastery of popular histories (*haishi* 稗史). No matter how suspiciously similar they might appear, if we take a step back for a moment, [we can see] it is simply that Enchō's spoken words are deeply draped in the essence of human emotions. The lowly and despicable hacks of the world who pander to woman and children strive to capture the condition of human emotions, yet only capture its surface in words that come across as lifeless. No doubt they will read the words of *The Peony Lantern* and be ashamed next to Enchō. I have written this preface by merely touching upon a few of my feelings on the subject.[17]

Shōyō identifies a phonocentric conceit behind the multiple layers of the ghost story that one might here, too, call "the ghost in the machine." In rapid succession, Enchō's breath is transmuted into verbal enunciation, recorded in shorthand, then turned into mixed kanji-kana script printed typographically. While he claims the spontaneity of the performance itself enables Enchō to rival even the illustrious Shunsui and Samba, it is shorthand that allows the literally and affectively moving words to manifest their full emotional force.

The strongest indication of Shōyō's support of shorthand as a medium for literature is expressed in his little-known essay "What Is Beauty?" published in the fourth volume of *Gakugei Zasshi* 学芸雑誌 (Academic journal, 1886). Foregrounding shorthand in relation to the discourse of

---

17. Shōyō, "Preface" to Enchō, *Kaidan botan dōrō*, 3. Translation is my own.

photography, he reiterates that shorthand may provide a potential foundation for so-called mimetic realism in literary texts.

> When we consider mimetic "beauty," we can see that indirect modes of painting have fallen into disuse, while the art known as filming (*satsuei no jutsu*) has become increasingly advanced. As the reader is surely well aware, in recent years the art of filming has progressed to the point where not only can we create a copy of an entire body, but also render details of clothing in perspective just as they appear. No matter how ingenious a painting is, it cannot compare with a photograph (*shashin*). In writing, too, there is what is known as shorthand notation (*sokkihō*), which is a method of capturing words just as they are (*kotoba wo ari no mama ni utsushitoru*). . . . It may be a critical judgment, but this is not extremely perfect novel writing. Nevertheless, a true connoisseur who sees this photographic writing ought not reject it for being written in shorthand. Rather, for those who delight in fanciful popular histories more cleverly conceived than Samba's disciples, and for those who see beauty in skillful artisanship, the two words "mimetic photography" (*moyō shashin*) should not be disparaged, but instead provide the basis for a solid foundation.[18]

How striking that in the midst of the debates over language and script reform in Japan in the 1880s, the preeminent literary theorist, Shakespearean translator, and English literature scholar found room to praise the decidedly mechanical art of shorthand! This endorsement of a photographic model of realism was a decisive break with Fenollosa's aesthetics of fine art and his insistence upon the value of beauty in art before truth.

Shōyō's endorsement of the language of *rakugo* and his praise for Enchō's skillful portrayal of sentiment underscores the often-overlooked fact that while *The Peony Lantern* combines two of the four major *rakugo* genres, namely *ninjō banashi* (sentimental stories) and *kaidan banashi* (ghost stories), it was exclusively singled out by Wakabayashi, Shōyō, and others for the attributes of sentiment. Later scholars similarly took up Shōyō's advocacy of psychological realism and found it convenient mostly to ignore elements of the supernatural in the text.

18. Tsubouchi, "*Bi to wa nani zo ya*," 219.

Enchō, on the other hand, was a more ambivalent co-conspirator in the death of the ghost story, or at least its sublimation beneath the discourses of scientific truth and psychology. His first *rakugo* performance, *Kaidan kasane ga fuchi* (1858), was transcribed some twenty-eight years later by Wakabayashi and renamed *Shinkei Kasane ga Fuchi* 真景累ケ淵 (True view of the Kasane Pool, 1886). Gerald Figal calls our attention to the introduction of this text, which, like *The Peony Lantern*, offers a preamble to better situate its textual reproduction. In his opening remarks, Enchō plays on the homophony of "*shinkei* (nerves), the word that had become fashionable by mid-Meiji to refer to forms of mental (nervous) disorder (*shinkeibyō*). Thus the 'true view' of the supposed supernatural events recounted in the tale about Kasane Pool is that they are the product of the protagonist's nervous disorder and not of an otherworldly visitation."[19] The "true view" is no longer visible to the eyes as a spectral presence, but directed by the nerves as a psychological malady. While modern civilization sought to banish the irrational or otherworldly, Enchō would continue to mine the public's love of ghost stories until the end of his career.

The Transparency of the Novel

I want to turn now to a critical analysis of how the concepts and practices of shorthand, as well as the ongoing changes in the verbal, visual, and print culture regimes of early Meiji, were effectuated in Futabatei's *Floating Clouds* and Shōyō's *Essence of the Novel*.

In *Kansei no henkaku* 感性の変革 (*Transformations of Sensibility*, 1978–1982), Kamei Hideo lays claim to a distinctive feature of *Floating Clouds* that he calls "the non-person narrator." This narrative position is distinct from the "I-ness" (*watakushi-sei*) or intentionality of the author, or for that matter, the protagonist. Considering that Kamei does not provide a comparative framework or historical basis for its applicability beyond a select few Meiji authors, we must ask from the outset whether it should be understood as a unique product of Japanese literary modernity

19. Figal, *Civilization and Monsters*, 27–28.

or has more universal applications. Perhaps it is better to call this idiosyncratic figure the "non-*persona* narrator" insofar as it lacks identifiable marks as a character or individual. In any event, the anonymity afforded this figure cannot conceal that its use of language and privileged gaze are decidedly male. Consistent with our earlier examinations of Anderson's thesis on the imagined community, a new narrative voice emerges in the vernacular texts of newspapers, novels, and other media that sustained a horizontal fraternity of the modern nation-state. As Miyako Inoue argues with regard to the emergence of this sort of narrator's class and gender affiliations, "I say 'him' because this narrator, this citizen, was presumed to be (the middle-class) male, and he alone had full and legitimate access to the newly emerging liberal-democratic public sphere."[20]

In his reading of *Floating Clouds* in *Transformations of Sensibility*, Kamei resists any connection to the shorthand transcription of *rakugo* by attempting to forestall claims made as early as 1958 by literary scholar Terada Tōru.[21] Rather than trace a limit outside the text that would disclose any nonliterary origins, Kamei compares *Floating Clouds* with *kanbun fūzokushi*, an earlier and unquestionably *literary* genre of manners and customs. Moreover, while he is quick to acknowledge the *possibility* of dialogic contributions by Shōyō, he quickly shifts his analysis to narratological grounds where Futabatei's unitary authorship and the unitary voice of the narrator cannot be gainsaid.[22] In his later work *Meiji bungaku-shi* 明治文学史 (History of Meiji literature, 2000), Kamei proves only slightly more amenable to the literary history of shorthand. Nevertheless, he remains at best noncommittal in evaluating its claims to high fidelity and its role in shaping narrative voices and modern subjectivity.

Needless to say, my reading differs considerably from Kamei's. I submit that the "non-person narrator" may be nothing more than an

---

20. Inoue, "Gender, Language and Modernity," 401.

21. Kamei, *Transformations of Sensibility*, 9, 15. See also Terada's article, "Kindai bungaku to nihongo," 1–27.

22. There is a double movement in Kamei's argument where he denies pursuing authorial intentionality, only to claim that this unique narratological stance could only have come from the persona of Futabatei: "all the similarities with Shōyō notwithstanding, there is one thing we would never find in Shōyō's writing that we find in *Ukigumo*—its narrator, and his somewhat discomfiting moments of self-exhibition" (*Transformations of Sensibility*, 99).

approximation of the spontaneity of *rakugo* storytelling (*hanashi* 噺), in which the raconteur darts in and out of various dramatic personae to entertain, or even dazzle, the audience. Such effects in a live show may be as simple as a change in gesture, facial expression, or intonation. I would offer as an example in *Floating Clouds* the scene that immediately follows the well-known comic disquisition on men's beards and suits that begins the novel. The beard scene is already widely recognized as employing *rakugo* techniques, but there is perhaps an even more revealing episode shortly after Bunzō exits the hustle-bustle of the city streets outside the Kanda gate. As he enters his aunt's house, where he lives, the unnamed narrator invites the reader: "Shall we go in, too?" (*isshyo ni haitte miyō*).[23]

Bunzō ascends to his spartan six-tatami-mat room, changes from suit to kimono, and sits down at his writing desk. The maid hands him a letter from his mother just delivered from the country, but he cannot read it due to the maid's interfering chatter. Seemingly in the same breath, she fills his ears with sighs of praise for his cousin Osei's beautiful outfits and complains about his aunt's cruel comments about her homeliness. Protesting the aunt's acid-tongued remark that the maid's use of makeup was as pointless as frost landing on a lump of charcoal, she cries to Bunzō, "Isn't that going too far? Don't you think that's going too far? No matter how plain I am?" (*Anmari jyā arimasen ka? Nē—anata nanbo watakushi ga bukiyō datte anmari jya arimasen ka*).[24] The narrator not only captures her speech, but also her mannerisms, such as excitedly wetting her lips and "making figures with her hands as she speaks" (*te de katachi wo koshiraete mise*).[25] In short, the verbal portion of the text is rife with elements one might expect in a theatrical performance.

This is not the end of the scene's representation in the narrative, however. The first of several illustrations scattered throughout the novel mirrors this passage (figure 8.3). An obvious discrepancy in Bunzō's attire notwithstanding—he remains in the Western suit—it opens a visual window that parallels and reinforces the narrative voice. For instance, a small text box (or "floating cloud") in the upper left-hand corner of the

23. Kōno and Nakamura, eds., *Futabatei Shimei zenshū*, 1:10.

24. Ibid., 1:12. In English, see Ryan's *Japan's First Modern Novel: Ukigumo*, 201. The kanji are glossed with kana whose pronunciations are given here in Romanization.

25. Kōno and Nakamura, eds., *Futabatei Shimei zenshū*, 1:12.

8.3 Illustration from *Ukigumo* (Floating clouds), 1886. Photograph courtesy of National Diet Library.

picture plane reproduces the speech quoted above in the cursive style of Edo popular fiction. The illustrations are an inconvenient presence in the text for scholars who endorse the notion that Futabatei realized one of the main tenets of realism in *Essence of the Novel*—namely, the elimination of pictorialism. Instead, the illustration restages the enunciatory act, disrupting the logic of a purely verbal or homogeneously typographic text. Curiously enough, it also looks a bit like something we might readily find off-stage at the *yose* theater: the raconteur delivering his stories to a shorthand reporter, who transcribes them onto paper. The plain-looking and plain-speaking maid—her mode of speech is quintessential Tokyo dialect—sits just as a raconteur would when performing with lips, fingers, and eyebrows aflutter. Bunzō, meanwhile, peevishly sits at his desk with a collection of writing brushes, ink stone, and the partially unraveled letter in his hands, craning his head to listen. Collectively, these details call into question whether something as obvious to contemporaneous readers as the legacy of *rakugo* transcription was not being signaled by writer and illustrator in tandem at the outset of the novel.

In "The Origins of My Unified Style," Futabatei makes the "confession" (*zange*) that his discovery of the unified style came about as he struggled against the impasse of classical writing styles. When he sought Shōyō's advice, the more experienced writer encouraged him to pattern his narratives after Enchō's *rakugo*. The result, he somewhat immodestly points out with the benefit of hindsight when the unified style was already in ascent, was as follows: "I imitated Master Enchō to signal [the start of] a new Japanese style" (*Enchō-shi no monomane de Nihon no shin-bunshō no kōshi*). By contrast, Yamada Bimyō, whose own use of the unified style in *Fūkin shirabe issetsu* 風琴調一節 (A tune on the accordion, 1887) preceded *Floating Clouds*, makes explicit the shorthand connection and debt to *rakugo*: "in a word, it should be like the transcription of Enchō's sentimental stories, albeit with embellishment" (*hitokuchi ni ieba, Enchō-shi no ninjō-banashi no hikki ni shūshoku wo kuwaeta yō na mono*).[26]

Although he was born in Edo, Futabatei was raised and educated until college in Nagoya. His relationship to the Fukagawa dialect by no means came naturally to him, but was the result of studied effort.[27] In "The Origins of My Unified Style," Futabatei professes to have argued with Shōyō over the use of classical epithets and flowery language (*bi-bun*), which he did not want to use. Nevertheless, he deferred to many of Shōyō's requests on the basis that readers might otherwise reject the results. Although there is no direct mention of shorthand transcription in the essay, Shōyō's advice quoted by Futabatei positively rings out with the imperative to write things down just as they are: "Just like that! Don't change a thing and leave it raw and untouched, just as it is" (*kore de ii, sono mama de ii, nama jikka naoshitari nanzo senu hō ga ii*).[28] As I have already noted, Futabatei, who was a translator of English and Russian,

---

26. Yamamoto, *Kindai buntai keisei shiryō shūsei*, 362.

27. See Ryan, *Japan's First Modern Novel: Ukigumo*, 80–85.

28. We might consider this faithfulness to the immediacy of speech in *Ukigumo* alongside another novel of multiplied origins and high-fidelity recording published three years later: Morita Shiken's translation of Victor Hugo's crime journal *Things Seen* (1887) as *Tantei Yūberu* 探偵ユーベル (Detective Hubert, 1890). Renowned for his meticulous word-for-word substitutions of the original language of the text, down to the introduction of *kare* and *kanojo* as personal pronouns equivalent to "he" and "she,"

and a lay linguist,[29] also readily admitted to studying Samba's *Ukiyo-buro* to reconstruct the Fukagawa dialect.

Much as Futabatei's recognition of shorthand transcription threatens to diminish the belles-lettristic value of *Floating Clouds* by associating it with the mechanical rather than liberal arts, there is the problem of multiple authorship further undermining the Romantic conceit of the author as an individual genius. In keeping with the longstanding custom of using a more established author's reputation to sell untested new authors, the first of the novel's three parts was published in Shōyō's name. The second part lists Shōyō and Futabatei as coauthors, while only the last installment credits Futabatei exclusively. Needless to say, this omits the illustrator whose work I mentioned earlier: celebrated woodblock print artist Yoshitoshi Taisō. *Floating Clouds* was very much a product of the verbal, visual, and print regimes in mid-transformation in the 1880s, and its dialogic production underscores this fact.

Finally I want to turn to Tsubouchi Shōyō's *Essence of the Novel* for a discussion of transcriptive realism and the literary hermeneutics that sustain its effects. Despite its iconic status as the literary tract that launched a thousand novels, *Essence of the Novel* also has the unfortunate distinction of being among the most maligned and misunderstood texts in formation of the canon. It is almost de rigeur in postwar scholarship to pigeonhole Shōyō's knowledge of literature, the extent to which he failed to live up to his own ideals, and of course, the notion that he facilitated Futabatei while slowly disappearing from the literary scene as a novelist

---

Morita points toward another aspect of verisimilitude and realism in the production of the modern Japanese novel.

29. Futabatei Shimei's article *"Esuperanto kōgi"* エスペラント講義 (Esperanto lectures) in the September–October 1906 issue of *Gakusei Times* was published to coincide with the founding of the first Esperanto Association in Japan. Futabatei makes no mention in the article of another significant historical coincidence, namely that his novel *Floating Clouds* and Esperanto were published in 1887. Certainly L. L. Zamenhof's creation of a planned world language and Futabatei's literary debut coincided within a larger discourse of national and international language reform. In the article, Futabatei reminds the reader of the absence of a single world language by pointing out there was only English (or pidgin-English) for commerce and French for diplomacy. While Futabatei did not rush to judgment on the future success of Esperanto, the seriousness with which he took it accurately reflects the degree to which global linguistic change seemed possible well into the early twentieth century.

in his own right. Naturally, these critiques fail to do justice to his tremendous output in literary and theatrical reform, education, and translation. Although a proper reply to these charges against him are no doubt long overdue, here I simply wish to demonstrate how the notion of transcriptive realism combined with a new literary hermeneutics in his most important and widely read theoretical tract.

The negative appraisals of Shōyō are concisely summarized in Atsuko Ueda's "The Production of Literature and the Effaced Realm of the Political," which maintains that Shōyō's *Essence of the Novel* was pivotal in the suppression of explicit political discourse from the takeoff of the modern Japanese novel.[30] Affirming the popular view that *Essence of the Novel* was the starting point of canonical literary discourse, Ueda nevertheless takes issue with its privileged status, claiming that "literary history already restricts its boundary to the predetermined discursive region of literature, which was, in large part, shaped by the *Shōsetsu shinzui* cliché."[31] This formulation, she argues, began with Shōyō's depoliticization of literature and movement away from the concerns of the political novel. Picking up from my discussion of Yano's *Illustrious Statesmen* in chapter 7, I would argue that what was effaced from the production of literature was not politics or ideology per se, but the new horizon of indexicality conveyed by shorthand and the changing verbal, visual, and print culture regimes in the middle of the Meiji era.

Shōyō's literary theory took its point of departure from Fenollosa's aesthetics of fine art. It is sometimes forgotten that the pseudonym Shōyō 逍遥 literally means "peripatetic" and was derived from Meiji-era Japanese translations of Aristotle, one of the philosophers whom he studied with Fenollosa. It was a nom du plume consistent with their mutual, if rarely concordant, *quest*-ioning of philosophical categories of knowledge.[32] In his opening salvo in *Essence of the Novel*, Shōyō articulates his view of literature over and against Fenollosa's aesthetics. Indeed, not only

---

30.   Ueda, "The Production of Literature and the Effaced Realm of the Political," 61–63. She provides a useful overview of some of the major postwar scholarship on *Essence of the Novel* by Nakamura Mitsuo, Ino Kenji, Maeda Ai, Kamei Hideo, Donald Keene, and Marleigh Grayer Ryan.

31.   Ibid., 63. The term was coined by literary scholar Nakayama Akihiko.

32.   There is a curious resonance here with Fenollosa, who, when he was made an honorary member of the Kano school of painting in 1884, adopted the artistic name

does Shōyō declare in the section entitled "Benefits of the Novel" (*shōsetsu no hieki*) that "the novel is fine art" (*shōsetsu wa bijutsu nari*); he also follows Fenollosa's precedent in denigrating woodblock prints and praising oil painting.[33] In one instance, he sees differences in contemporary Japanese and European literature as tantamount to the difference between the patrician and plebian visual arts of the Edo period: "Comparing the Japanese novel with its Western counterpart is thus like comparing the Utagawa *ukiyo-e* woodblock prints with Kano paintings. The prints, though certainly not clumsily executed, lack the quality of refinement, having nothing to offer the viewer's aesthetic sensibilities."[34] Or again, in a separate passage he insists: "the novels prized in Japan are crude; they lack the qualities of art. They occupy a position like that of *ukiyo-e*, which cannot be called genuine painting."[35]

These descriptions are used to contrast ennobling spiritual or inner beauty with vulgar truth. This opposition also obtains between the epic poetry and myths of the past and the modern novel. Whereas the former elevates beauty above all else, the latter concerns itself with the accurate portrayal of contemporary human emotion, behavior, and psychology. Shōyō demonstrates a keen grasp of the novel as a class- and gender-based instrument of social change, noting its capacity to inspire as well as to reflect lived social realities. It was this attention to social problems that prompted his famous denunciations of the didacticism of Edo *gesaku*, which, like the romance and novel in early modern Europe, were at best excessively moralizing and at worst paid a thin lip service as a cover for excessive licentiousness.

Notwithstanding the textual diversity of adaptations, translations, and political novels epitomized by Yano's *Illustrious Statesmen, Essence of*

Yeitan 永探, or "Eternal Search." See Chisolm, *Fenollosa: The Far East and American Culture,* 55.

33. In his later career as curator of the Japanese art collections at the Boston Museum of Fine Art and as author of *Masters of Ukiyoe* (1896) and *Epochs of Chinese and Japanese Art* (1912), Fenollosa became something of a spokesman for the beauty and historical value of woodblock prints. In "The Truth of Art" and his earlier works during his time in Japan, however, Fenollosa was harshly critical of their vulgarity and dismissed the faddishness of Japonisme by Europeans unschooled in the Tosa and Kano schools of painting.

34. Shōyō, *Essence of the Novel,* 36.

35. Ibid., 38.

the Novel was the first sustained attempt in literary theory to analyze Japanese and Western literature together. Shōyō wove together strands from nineteenth-century *kokugaku* studies, including Motoori Norinaga's aesthetics of *aware* and his considerable knowledge of *gesaku*, as well as major works from the English canon. Akin to Yano's denouncing "popular histories," Shōyō contrasted the modern novel in both Japanese and English forms with premodern Japanese romance. While he regarded the romance as fanciful and as having no basis in reality, Shōyō insisted that the novel had achieved the apex of affective response in contemporary times. In his famous elaboration upon the purpose of the novel, Shōyō argues, "the *shōsestu*, by which I mean the novel, takes a different tack, with the portrayal of human feeling and customs in the world as its central aim" (*shōsestu sunawachi noberu ni itarite wa kore to kotonari, yo no ninjō to fūzoku o ba utsusu o mote shunō to nashi*).[36] Of course, it was this register of affect that made the novel a site of poeisis, a subjective technology for bringing about the transformation of social relations.

Shōyō's prescriptive writings on psychological realism and mimeticism demonstrate how he imported concepts from the visual discourse of *shashin* and the conceptual vocabulary of shorthand into literary theory. He criticizes Kyokutei Bakin's moralizing *yomihon Nansō Satomi Hakkenden* 南總里見八犬伝 (Tale of eight dogs, 1814–1842) for its excessively allegorical narrative that lacks verisimilitude to any recognizable human being. "These eight 'wise'-men were Kyokutei Bakin's ideal characters, not photographs of the men of his age, which explains the disparity" (*hakkenshi wa Kyokutei Bakin no aidiaru no jinbutsu ni shite, gensei no ningen no shashin ni araneba, kono futsugō mo ari keru naru*).[37] This metaphorical use of photography as "true portrayal" is reinforced by his conviction that the author "should not interpose his own designs on human feeling, whether for good or ill, but copy things just as they are, as an observer" (*aete onore no ishō wo mote zenaku-jasei no jyōkan wo tsukuri-mōkuru koto wo ba nasazu, tada bōkan shite ari no mama ni mosha suru kokoroe ni te aru beki nari*).[38] Portraying the novel as a kind of chess game that the author is observing rather than playing, he recommends a mode of descrip-

---

36.   Takenouchi, ed., *Nihon bungaku zenshū*, 1:106. Translation is my own.
37.   Ibid., 1:119. Translation is my own.
38.   Ibid., 1:119–20. Translation is my own.

tion that will not interfere with the outcome. Indirectly invoking the compositional imperative of shorthand, he polemically reiterates, "When one copies things just as they are, for the first time that can be called a novel" (*tada ari no mama ni utsushite koso hajimete shōsetsu to mo iwaruru nare*).[39]

It is against this backdrop that I find especially problematic Ueda's contention that Shōyō did not articulate a clear sense of realism, but only used it in a nebulous opposition to didacticism. She insists:

> Shōyō was not calling for "mosha" or "ari no mama" as mimetic realism. We need to extract the textually specific meaning of terms in the text, provisionally treating terms as empty signs and finding what certain terms align themselves with or are defined against within the text itself. Shōyō often posits the key constituents of modern fiction through a chain of negation. For that reason, I wish to focus not so much on what he positively identifies as defining characteristics of fiction, but on what he negates as that which is not.[40]

Arguing that the only references to mimetic realism occur when the intentionality of the author is opposed to didacticism, she continues, "In fact, throughout the text, the meaning of ari no mama is never positively identified, only posited through such negation."[41] To treat these critical concepts as "empty terms" or mere "negations" is to forgo any attempt at understanding the media history, as well as literary and philosophical parameters that structured his thought.

Ueda derives her rationale in part from Maeda Ai's essay "Modern Literature and the World of Printing," a key passage of which she quotes in her text and which I will reproduce here in full:

> It is commonly believed that *Essence of the Novel* by Tsubouchi Shōyō valorized the earliest form of a concept that corresponds to realism in modern literature. What Shōyō actually attempted to develop in *Essence of the Novel*, however, was a problematic broader than anything that can be contained within the framework of realism. Take the term *mosha* [copy], for example. We tend to understand the term as being interchangeable with

39. Ibid., 1:120.
40. Ueda, "The Production of Literature and the Effaced Realm of the Political," 69.
41. Ibid., 70.

shajitsu [realism], but in Shōyō's usage it seems to have been closer to a term used by Edo painters to mean the affixation of their names to their paintings; in other words, the characters for mosha, when following a proper name, meant painted by so-and-so. When we discuss literature today we have at hand a received system of literary terms, but that was not the case for Shōyō, who had to create the very terms he used to discuss literature. We must be mindful of that situation when we read him.[42]

While Maeda was responsible for many groundbreaking studies of Edo and Meiji visual culture, he never introduced the shorthand connection in his scholarship, generally ascribing the conceptualization of realism to literary theory, reader response theory, and more broadly, the transformations in print culture. Of course, Maeda was among the first scholars in Japanese literary studies to challenge disciplinary lines and investigate the transformations of print culture and writing technologies from Edo to Meiji. While Maeda makes passing reference to the transcription of *rakugo*, he never directly pointed to its role as "verbal photography" that transformed the concepts and practices of the modern novel. This factor severely impinged upon his ability to explain Shōyō's quite specific use, and media-specific grounding, of terms such as "copying," "photography," and "realism."

While Shōyō discusses at length relations of orality, literacy, and visuality, there is no explicit statement about shorthand transcription in *Essence on the Novel*, only a few suggestive phrases and mentions of realism. One plausible explanation for the absence of shorthand is indirectly supplied by Peter Kornicki, who argues that *Essence of the Novel* was written mostly between 1881 and 1883, prior to the advent of Japanese shorthand, Shōyō's involvement with *The Peony Lantern*, or his other writings on shorthand and realism:

> Opinions differ about the genesis of *Shōsetsu shinzui* and the details of its composition, but it is widely accepted that the first draft was complete by the end of 1881 at the latest and that by the end of 1883 something close to

42.   Ibid., 68. See also Maeda, *Text and the City*, 255–56. Despite her own attempts to strictly demarcate Shōyō's terminology, Ueda's claim for a tension between *ninjō*, *fūzoku*, and *setai* (sentiment, customs, and manners) on the one hand, and politics, on the other, overlooks how the former three categories are not only aesthetic but *poetic*: it is through the emotional register that they are capable of reordering social relations.

the present text was in existence. After some delay, it was published first in nine fascicles from September 1885 to April 1886, and then in May 1886 in a two-volume edition, with a second impression appearing in August 1887.[43]

Needless to say, the difference of one or two years utterly changes the implications for Shōyō's knowledge about shorthand notation. This is not to suggest the entire text was composed at the earliest possible date, or that Kornicki's dates might not also be open to revision. Still, the opening reference to Fenollosa and other comments such as his opinions on the Romanization Society and Kana Society, unambiguously locate key portions of text on a time line from 1882 to 1884.

On the related topic of realism in the theater prior to Shōyō, there is, parenthetically, but a single enigmatic critique of Danjurō IX and the reform of theater as *katsurekishi* and *zangirimono* by a raconteur whom Nanette Twine and others have assumed is Enchō.

> As a certain raconteur in Tokyo once pointed out, the taste of theater-goers has changed considerably with time. Realistic performances, no matter what their subject, are enthusiastically applauded. People praise the subdued colors of Ichikawa Danjurō's costumes for their restrained elegance. Stage dialogue resembles ordinary speech, and shouts of approval greet the substitution of theatricality for words to convey a meaning. In years to come, probably, Danjurō and his contemporaries will play heroes confined to bed by illness, and nap through several acts in the dressing room! Who can tell, the humorist asked, whether public opinion might not swing in favor of cropped hair and unpainted faces (*sugao zangiri*) on stage? He joked about how strange it is, and his words bear out my point.[44]

This passage belies canonical claims that *katsurekishi* and *zangiri* were universally panned by audiences. There is something odd, however, about this attribution to Enchō. Although he was hardly an eager accomplice in the ideological project of "Civilization and Enlightenment," he had no qualms about profiting from the celebrity brought to him by shorthand

---

43. Kornicki, *The Reform of Fiction in Meiji Japan*, 26.

44. Shōyō, *Essence of the Novel*, 20. In Japanese, see Shōyō Kyōkai, eds., *Shōyō senshū*, 3:37.

transcription. This extended to his dealings with the kabuki theater and its powerful actors. In fact, Enchō's *Eikoku kōshi no den* 英国孝子の伝 (Tale of a filial Englishman, 1896) was made into a kabuki play performed by Danjurō and renamed *Seiyō hanashi Nihon utsushi-e* 西洋噺日本写絵 (A Western tale in Japanese magic lantern pictures). It is a title rich in associations with the medial concepts and practices of *utsushi*, which means both copying and projection. Enchō attended the January 1886 performance at the Shin-tomiza in Tokyo with his favorite amanuensis, Wakabayashi Kanzō, in tow.[45] It seems unlikely, then, for Enchō to disparage Danjurō's work, which was, in a sense, also his own.

The difficulty in assessing Shōyō's views on language reform and the advancement of new forms of verisimilitude through experimental scripts and vernacular language is that he did not readily conform to any single ideological camp. In fact, he was not an advocate of the unified style any more than of romanization or kana-only script. On the highly contentious topic of literary language and language reform, Shōyō makes two crucial observations in *Essence of the Novel*. The first pertains to what he regards as the salutary improvement in newspaper serial fiction (*tsuzuki-banashi*) that had at the time begun to introduce Tokyo colloquial speech (Fukagawa dialect) into print, in part reflecting the city's recent name change and newfound pride as the imperial capital. The second comes in a passage that could only have been written after 1884, where he remarks about the recent establishment of the Kana Society and Romanization Society. It is evident that Shōyō was well informed, if not entirely convinced of the merits, of the international milieu of phonetic scripts and language reform. "I do not think that the supremacy of either the kana or the Roman script is really their ultimate aim. The long-range goal of these kindred spirits is to unify all countries into one vast republic with as far as possible a common political system, national language and customs" (*udai no bankoku wo itto shite ichidai kyōwakoku no arisama to nashi, oyobu beku dake fūzoku wo mo mata seitai wo mo kokugo wo mo dōitsu narasimemu to nozomu ni ari*).[46] If the objective of both groups was to

45. Fujikura, *Kotoba no shashin wo tore*, 246. After the performance (which included other illustrious figures such as Ichikawa Sadanji), Enchō and Wakabayashi paid their respects to Danjurō backstage.

46. Shōyō, *Essence of the Novel*, 73. In Japanese, see *Tsubouchi Shōyō senshū bessatsu*, 3:113. I have changed Twine's term "language" back to "national language," but otherwise retain her translation.

establish equivalence with the West, whose hegemony seemed impregnable, Shōyō argues, then the logical or perhaps inevitable choice would be to go the route of Romanization. Although he leaves open the possibility of a new *kusazōshi* style in the Roman script, he hastens to add that he endorses neither group and welcomes their clarification if his judgment is in error.

The essence of Shōyō's aesthetics boils down to a question of literary hermeneutics—that is, to his desire to read literary texts as invested or encoded with illusory depth. Shōyō describes how effortless reading and the imaginative powers of a transparent medium make literature superior to theater or the arts:

> Whereas the depiction of personality in the theater is limited by its dependence on visual and aural appeal, the novel, by communicating directly with the reader's mind and stimulating his imagination, has a much wider range. On stage nature, scenery, houses and furnishing are represented by painting or props. Mechanical devices are enlisted to give evidence of rain, wind and storm. All these things are described in the novel, captivating the reader's mind's eye [*shinmoku*]. The degree of interest aroused in him thus depends on the strength of his own imagination.[47]

The upper limit of what is possible thus depends on one's capacity for reading, which, like the presumed unity of the authorial hand-voice-mind, is responsible for re-creating the world of the text. Equally significant for Shōyō is the re-lationship of literacy and visuality. He reintroduces the concept of "the mind's eye," which we might note was previously used in his teacher Fenollosa's "The Truth of Art." In *Essence of the Novel* it refers to the visual aspect of literature accessed by the reader's imagination through the verbal text. Consistent with the early Fenollosa's insistence upon the separation of poetry and painting as per the aesthetics of Lessing's *Laocoön*, Shōyō calls for the removal of illustrations and the "pictographic" aspects of Meiji literature.[48] Of course, illustrated cover art, frontispieces, magazine illustrations, and the like remained integral to the literary culture of the era with the participation of famous artists

47. Ibid., 21.
48. For additional perspectives on this issue, see Inouye's "Pictocentrism" and Li's *Hyōshō kūkan no kindai*, 143.

such as Yoshitoshi and Kaburagi Kiyokata. Yet as soon as Meiji literary journals, newspapers, and magazines set visual elements apart and subordinated them to a secondary referential function (as in depicting a scene already described in words), the break with the existing visual-verbal regime was complete.

Of course, we must insist that readers never actually read word-for-word in the supposedly transparent medium. In the discourse network of late nineteenth-century Germany, as Kittler reminds us, Nietzsche sought to strip away the accumulated weight of over a century of Romanticism and expose the "stinking cadaver" of German poetry and philosophy sustained by hermeneutics. Not only was the "quintessence of the personality of their authors" suspect for Nietzsche, but also the notions of subjective interiority and imaginative clarity that springs into the mind of the reader from the words effortlessly deciphered on the page. In his fiery efforts to de(com)pose Romanticism, Nietzsche averred that the steadfast reader was nothing but a myth: there was at best only skimming, scanning, and guesswork from the supposedly faithful reader. Nietzsche decries these slovenly tendencies in *Beyond Good and Evil* from 1886:

> Just as little as a reader today reads all of the individual words (let alone syllables on a page)—rather he picks about five words at random out of twenty and "guesses" at the meaning that probably belongs to these five words—just as little do we see a tree exactly and completely with reference to leaves, twigs, color, and form; it is so very much easier for us simply to improvise some approximation of a tree. . . . All this means: basically and from time immemorial we are—accustomed to lying. Or to put it more virtuously and hypocritically, in short, more pleasantly: one is much more of an artist than one knows.[49]

The artist and the decay of lying—Nietzsche begins to sound positively Wildean. Yet the comparison is apt, and all the more insidious to Nietzsche's righteous indignation in the re-enfolding of the hermeneutic lie into the knowing conceit of art for art's sake. Oscar Wilde's "The Decay of Lying: An Observation" (1889) articulates the lack in nature, or the real world, for which the artifice of representation make amends:

49. Nietzsche, *Beyond Good and Evil*, 5:192, 105. See also Kittler's discussion of Nietzsche in *Discourse Networks 1800/1900*, 177–84.

When I look at a landscape I cannot help seeing all its defects. It is fortunate for us, however, that Nature is so imperfect, as otherwise we should have no art at all. Art is our spirited protest, our gallant attempt to teach Nature her proper place. As for the infinite variety of Nature, that is a pure myth. It is not to be found in Nature herself. It resides in the imagination, or fancy, or cultivated blindness of the man who looks at her.[50]

Independently or by surreptitious borrowing, Nietzsche and Wilde mutually indicate the paradox that resides in the reading, as well as composition, of the literary text.[51] On the one hand, the reader simply rushes along, sometimes stumbling through the thickets of meaning laid down in neat typographical lines. On the other hand, for Nietzsche and Wilde alike, Nature, too, is a false category, or rather a subset of Art. It is only by human endeavor that is consciously aware of its own artifice that the act of reading can overcome the *learned* blindness that would have us accept literary conceits as accurate portrayals of the natural, familiar, and innate.

By contrast, in Shōyō's disquisitions, the literary text transcends mere reflection or mirroring of the world by bringing about the communion of reader with the vision of the author. The author accordingly ought to not subjectively intercede in the narrative, but rather make his intentions known by and through the story. Unlike Nietzsche and Wilde, then, Shōyō explicitly endorses the Romantic conceit of author as a godlike Creator, with the totality of the world grasped through the transparent medium of the text.

The novel reveals what is hidden, defines what is indistinct, and brings together all man's innumerable passions within the covers of a book, thereby naturally stimulating the reader to introspection. There is a parallel to God, who created all things on earth but put nothing of himself into them, and authors with my way of thinking, who create a variety of characters with complete impartiality and present every aspect of daily life in a realistic manner.

50. Ellmann, ed., *The Artist as Critic*, 291.
51. On the mutual borrowings between Wilde and Nietzsche, see Ross, "Deceptive Picture," *New Yorker*, Aug. 8, 2011.

The realism of texts composed in this fashion reinforces the imagined community, the halluci-*national* identity that is not just an unfounded or mistaken perception of objects with no basis in reality—not unlike Marx's definition of ideology as false consciousness—but connected to the *subject-making* institutions of national language and literature.

Here we come full circle to the perceived capacity of shorthand to render speech into transparent writing. This is precisely what informs the concept of transcriptive realism: a "hallucinogenic" quality, as Kittler insistently calls it, that permits readers to see, hear, and above all, *feel*, what is not really there. Komori Yōichi identifies Japanese shorthand as an inscriptive system that translates across three levels of language: from speech to shorthand to mixed kanji-kana script. He observes that "even though the stenographer is physically present, he is erased (*shōkyo sarete iru*) so as to become a transparent linguistic medium (*tōmei naru gengo baikaisha*). Even today this is the illusion which enables it to function as a reproductive mechanism for voices to be heard from the printed word."[52] This is the projection (*utsushi*) that supplies the literal meaning of *shajitsushugi*, which I have translated throughout this book as "transcriptive realism" to underscore its media-historical origins. Of course, realism must always be construed in relation to some form of media or indexical grounds, so this polemical emphasis on transcription may be taken as a heuristic.

The hallucinatory experience Shōyō describes comes from the effortless reading of a transparent medium. It is fitting that the unified style, which has long been upheld for the transparency of reading, owes a great deal of its existence to shorthand, which more than any other form of nineteenth-century Japanese writing approached the condition of pure mediation by having its own presence effaced from the final text. To return to the formulation with which I began, the "real" ghost story that continues to haunt Meiji literature is the canonical origin myth that suppresses its own media history and other technical origins.

---

52. Komori, *Nihongo no kindai*, 125. It is important to clarify that Komori does not trace *ari no mama* back to media concepts or practices that I call attention to here, namely phonographic and photographic realism. He calls shorthand, oddly enough, a discursive product (*gensetsu shōhin*) that solicits an illusion of enlightenment and progress on the same order as Western languages. He does not deal with shorthand as a phonetic system, either. Rather, he sees it as a supplement to typographic media, which delivers a "verbal product" to the literary marketplace.

PART IV

The Limits of Realism

CHAPTER 9

Masaoka Shiki's "Scribblings"

What's first	*dai-ichi ni*
in the composition of lines	*sen no haigō*
then what comes next	*sono tsugi mo*
and next still after that—	*mata sono tsugi mo*
sketching, sketching!	*shasei, shasei nari*

—*Masaoka shiki zenshū*, vol. 6, 221. Translation is my own.

From the late 1880s, literary production proceeded apace with a diversity of literary styles and representations of Meiji life by Ozaki Kōyō, Yamada Bimyō, Higuchi Ichiyo, Kōda Rohan, and others. In the aftermath of *rakugo* transcription and nascence of the unified style, no literary figure so decisively shaped the direction of modern poetry and prose than Masaoka Shiki (1867–1902). In his capacity as a poet and critic, he invented the modern haiku from the worn-out constellations of traditional *haikai* poetics. Initially considered by some of his contemporaries a "failed novelist," he nevertheless created the prose style of "literary sketching" (*shaseibun*) that became the template for transparently phonetic, transcriptively realist novels. Although he would not live to see the fruits of his labors dominate first the literary mainstream, then achieve canonical status as the hallmarks of modern Japanese literature, Shiki's relentless experimentation and critiques of literary language provided an important transition from intermediary transcription of shorthand reporters to writers who would take matters into their own hands to "write things down just as they are."

For the title and first section of this chapter, I invoke as an organizing trope the title from Shiki's early collection of essays, *Fude Makase* 筆まかせ (Scribblings, 1884–1892),[1] to locate his work in the discursive

---

1. Shiki used several wordplays for "leaving it to the brush" based on different Chinese characters for the four volumes of *Scribblings*: 筆まか勢 for volume one, 筆任勢 for volume 2, and 筆まかせ for volumes three and four.

ruptures and peripatetic movements from shorthand transcription and artistic sketching to his application of statistics to mark the death by formalism of traditional Japanese poetics. In the second section I seek to account for Shiki's back-and-forth dialectics of life and literature, blood and ink. His tuberculosis manifested around 1895 and grew steadily worse, with a prolonged period of deterioration from 1898 to 1902. There is a Romantic conceit in excavating the archive upon which even Kittler depends—namely, that it will yield from its dusty crypts every secret thought and feeling of its authors and texts as a consistent and coherent system of meaning: the text of life. Shiki plays into such a conceit as a paragon of Romanticism on several levels. His unwavering obsession with writing things down, a mania that we might diagnose as *aufschreibesysteme*, only increased in intensity in spite, or perhaps because, of his physical debilitation. Shiki's writings continued unabated with an increasing awareness of the breath and fluids, as well as reprogramming of visual and verbal codes, that revitalized author and text.

Following Shiki's death in 1902, the principles of realism he advocated were widely adopted by major writers such as Kunikida Doppo, Shimazaki Tōson, Natsume Sōseki, and Tayama Katai and by lesser-known figures such as Nagatsuka Takashi, Itō Sachio, Terada Torahiko, Suzuki Miekichi, and Nogami Yaeko. This is not to say his ideas were uncontroversial or easily assimilated—time and again we see writers who, as it is conventionally described, struggled to define a narrative voice that was not alienated from the effects of representation or that indulged in an excess of artifice. *Heimen byōsha*, or "flat description," is a case in point, as Gerald Figal explains: "For Tayama, a truthful writing inscribed the responses of the five senses to the surrounding empirical reality. It was therefore preoccupied with surfaces, externals (hence, '*heimen*' *byōsha*). To introduce thoughts and feelings—that is, the imagination—of the writer would be to broach the 'naturally' sensed surface of things and permit the possibility of a falsifying fabrication."[2] While Katai lightens the narrative of a Meiji novelist wallowing in male self-pity in *The Quilt* 布団 (*Futon*, 1907) with occasional scenes of maudlin behavior that borders on slapstick, a far more effective critique and extension of literary sketching far beyond its original parameters occurs in Sōseki's *I Am a Cat*, which I will discuss in

---

2. Figal, *Civilization and Monsters*, 123.

the final chapter of this book. Although it is rarely recognized in relation to shorthand, I would argue that realism as transcription, indivisible from the unified style, marks the foundation of the modern Japanese novel.

Yanagita Kunio, the founder of Japanese folklore studies, is likewise deeply implicated in this regard. Yanagita's preface to *Tōno Monogatari* (Tales of Tōno, 1910) was a reformulation of the shorthand imperative for the purposes of ethnographic recording. In yet another example of what Marilyn Ivy calls a "parasitic preface" framing the text, Yanagita explains how he served as a phonographic amanuensis to Tōno native Sasaki Kyōseki, whom he describes as a "poor storyteller, but a man possessed with genuine sincerity" (*hanashi jōzu ni wa hizaredomo seijitsu naru hito nari*).[3] Apropos of the discourse networks of the post that linked the imperial center to domestic hinterlands and thus collapsed the distance between modernity in flux and a supposedly timeless folk tradition, Figal observes that Sasaki "has a reliability akin to that of a mail pony, strong but dumb, delivering messages intact from person to person over vast distances."[4] Yanagita continues with his characterization of how he recorded Sasaki's stories as follows: "I wrote them down just as I felt without changing a single word or phrase" (*jibun mo mata ichiji ikku wo mo kagen sezu, kanjitaru mama o kakitari*).[5] In this deft sleight of hand, Yanagita replaced the media capture of actuality with an affective grasp of what was intended, a folk essence.

In the slightly earlier essay "Genbun no kyori" 言文の距離 (The distance between speech and writing, 1909), Yanagita sought to make a clean break with the objective, observational stance of shorthand and literary sketching by emphasizing the powers of rhetoric, especially in folk traditions, over the presumed rationality and verisimilitude of the modern narrative voice and gaze. As Ivy explains, "Literature could only take place as truly literary by keeping its distance from the chaos of Japanese speech, as well as from the pretensions of a photographic and phonographic reproduction of the world (he writes of Japanese naturalist writers as bad amateur photographers)."[6]

3. Yanagita, *Tōno monogatari*, 55. Translation is my own.
4. Figal, *Civilization and Monsters*, 107.
5. Yanagita, *Tōno monogatari*, 55.
6. Ivy, *Discourses of the Vanishing*, 78–79.

The ethnographer remains a recorder and transmitter of native voice, or even the imagined voices of the past. But those voices are put down onto paper in the unified style and with the disciplinary mechanism of *minzokugaku* (ethnology) locking them into the framework of the modern nation-state. For all Yanagita's arguments to the contrary, even his approach subjected the common folk to a new kind of standardizing measure derived from transcriptive objectivity.

The Statistical Death of Japanese Poetry

In the first volume of *Scribblings*, Shiki wrote several short essays and fragments that indicate his critical interest in Enchō, *rakugo*, and somewhat cryptically, shorthand transcription. Shiki may well have gone to see Enchō perform live with his classmate Natsume Kinnosuke,[7] with whom he shared a passion for *yose*. By his own account, it was his early aspirations to follow in the footsteps of Shōyō, Futabatei, and Bimyō that in turn led him to the *yose* stage. He describes the various oratorical techniques used in *rakugo* in "Enchō no hanashi" 円朝の話 (Enchō's stories, 1890), and heeds the precedent of studying Enchō, or more precisely the transcribed effects of his performances, for the purpose of his own fiction writing: "Hearing these [oratorical] effects set me on the path toward writing novels" (*Kokora no guai o kikite yo wa shōsetsu no shukō mo kaku koso arita keredo satori tari*).[8]

Shiki's verse-form essay "Rakugo renzumō" 落語連相撲 (Rakugo sumo wrestling) provides a more precise analysis of his observations of *rakugo* transcription. Here, too, he describes the blurring of dramatic personae when watching a *rakugo* performance such that the audience is unable to tell where the characters leave off and the raconteur begins. Yet he was also mindful of the shorthand reporter behind the scene of writing who transformed the spoken word into literary text.

---

7.   Akio Bin calculates that if Shiki saw Enchō, it must have been at the peak of his popularity when Enchō was fifty-one years old and only two years before his retirement (*Shiki no kindai*, 76).

8.   Masaoka, *Shiki zenshū* (hereafter *SZ*), 10:91–92.

Gracefully, unsmilingly, they don't miss a single word or phrase, or make
  unnecessary chatter
The words written down just as they are spoken become Japanese literature
Unified style writers they are not
Ah, what fearsome sharp-tongues
Truly beyond comparison.

Odayaka ni shite, niyakezu, koto ni ichigen ichigo mo iiayamaru koto naku,
mudaguchi mo kikazaru tokoro wa hanashi sono mama no hikki o shite Ni-
hon no bungaku to shite. Genbun itchika o shite ironakarashimu anaosoroshi
no zeppō ya na masataru koto muron ni haberu.[9]

As this passage indicates, Shiki was critical of early efforts at the unified
style, which he saw as undisciplined, verbose, and unwieldy. This was a
view shared by many novelists of his generation, such as Ozaki Kōyō,
whose own experiments with the unified style began after the 1890s. By
comparison, Enchō and other professional storytellers were deft and
sharp-tongued, and their words, once transcribed, made for a real Japanese
literature. Shiki recognized that Enchō's novelistic style was a product of
shorthand. In "Genbun itchi no rigai" 言文一致の利害 (Advantages and
disadvantages of the unified style), which also appears in the first volume
of *Scribblings*, Shiki criticizes the prolixity of the unified style, but also
recognizes that writing, "like the transcription of Enchō's speech" (*Enchō*
no hanashi no hikki no yō ni), offered the best results for a written vernacular.[10] On at least one occasion, Shiki became the amanuensis to the
haiku gatherings he organized in 1897. As Mark Morris notes, "It was
apparently Shiki who at one early session, faced with a turn as recorder
of the discussion, decided to capture as well as he could a verbatim
account of the proceedings."[11] Nevertheless, Shiki never endorsed the
unified style as an exclusive medium for Japanese national language,
literature, or poetry.[12] Nor was he involved in any meaningful sense in

9. Ibid., 10:101. Translation is my own.
10. Ibid., 10:145. Translation is my own.
11. Morris, "Buson and Shiki, Part II," 258. Morris does not cite a specific passage,
only the title of the text, *Buson kushū kōgi* (Lessons on Buson's collected verses).
12. It is also likely that he was endorsing a more widespread view of the nadir of
the unified style in the early 1890s, when Ozaki Kōyō was popularly attributed with

the fixing of national language or script. He was concerned with formal experimentation with literary, poetic modes and their doubled refraction of life (increasingly focused on his own) and art. His legacy to transcriptive realism and the unified style were to a large degree incidental to his aims.

Shiki's reading of literary theory and texts in the 1880s included Shōyō's *Essence of the Novel, Tōsei shosei katagi* 当世書生氣質 (The character of today's students, 1885–86) and Yano's *Illustrious Statesmen.*[13] The model for his sole attempt at novel writing, however, was his contemporary Kōda Rohan's *Fūryūbutsu* 風流仏 (The Buddha of art, 1889). In 1892, Shiki wrote the Buddhist-inflected fantasy *Tsuki no miyako* 月の都 (Capital of the moon), which he presented to Rohan to solicit his approval. Finding little that was praiseworthy in terms of plot or narrative, Rohan instead diplomatically praised the *hokku* verses in the novel; thereafter Shiki turned away from fiction writing toward the composition and theorization of Japanese poetry. It was thus in a much different register of reinvigorating language that Shiki began experimentation with the prose vignettes he dubbed "literary sketching," and there was nothing inevitable about his discovery of it that would lead to its adoption as a mainstream practice by writers across the Meiji literary spectrum. Perhaps Shiki did not live long enough, nor was he sufficiently invested in the narratological potential of the style he pioneered to see it through to that extent. It would fall instead to Sōseki to fully exploit, in no small part through parody, the possibilities and limits of literary sketching in *I Am a Cat*, the first installment of which was published in Shiki's journal *Hototogisu* ホトトギス (Nightingale).

Insofar as everything for Shiki begins and ends with poetry, let us return to his interventions in the field of *haikai* poetics. Shiki transformed *haikai no renga* into haiku by taking poetic exchange and making it into individual literary production. This is not to say that Shiki shunned collaboration, but rather that he redefined its parameters. Among other activities, he experimented in *haiga* (haiku and painting) with artist Nakamura Fusetsu and worked with fellow poets to compose haiku

---

reviving a style he had originally helped to vanquish with his more popular *gazoku-setchū* (mixed ornate-colloquial) style.

13. See Beichman, *Masaoka Shiki*, 17–18.

in tandem. Nor was Shiki a polemicist of the separation of words and images along the lines of Fenollosa, Koyama, or Shōyō. Despite his vociferous attacks against what he saw as conventional *haikai* thinking and its forms of faux aristocratic communality and self-enclosed erudition, Shiki's poetics did not seek to abolish the form of traditional haiku. The rhythmic "cutting" syllables such as *ya*, *kana*, and *keri* remain, as does the notion of a unifying theme (*dai*).

In *Scribblings* Shiki reveals that some of his early cues on writing came from Herbert Spencer's *Philosophy of Style* (1871), which he read in translation as a university student in 1888—a rather profound irony, or perhaps confirmation, for those who see Shiki as the quintessential Naturalist before Naturalism. Nevertheless, Spencer's text has some surprises of its own—notably that it begins with a quotation from Laurence Sterne's *Tristram Shandy* (1760) to the effect that good argumentation does not require formal training or even correct grammar.

> Commenting on the seeming incongruity between his father's argumentative powers and his ignorance of formal logic, Tristram Shandy says:—It was a matter of just wonder with my worthy tutor, and two or three fellows of that learned society, that a man who knew not so much as the names of his tools, should be able to work after that fashion with them." Sterne's intended implication that a knowledge of the principles of reasoning neither makes, nor is essential to, a good reasoner, is doubtless true. Thus, too, is it with grammar.[14]

This is precisely the sort of point that language ideologues and social Darwinists such as Mori and Isawa, who sought to fix social relations by means of orthography, failed to grasp. Spencer's conception of language is consistent with the broad regime change witnessed in the nineteenth century as eloquence was first captured, then gradually superseded, by media of recording and transmission. Spencer argues: "Regarding language as an apparatus of symbols for the conveyance of thought, we may say that, as in a mechanical apparatus, the more simple and the better arranged its parts, the greater will be the effect produced."[15] Spencer almost seems to

14.   Spencer, *Philosophy of Style*, 9.
15.   Ibid., 11.

anticipate the manufacturing efficiencies of Taylor and Ford, which are, at their root about standardization of best practices and eliminating waste.

Matsui Takako points out that Shiki adopted many of the same principles in his poetics based on the concept of the "minor image," which Spencer usefully sums up under the heading "Arrangement of Minor Images in Building Up a Thought."[16] It is remarkably refreshing to think that Shiki's concepts of brevity and concision derived not from the purportedly minimalist essence of Japanese poetry or language, but from Spencer. It became for Shiki one of the early marshaling grounds for his reappraisals of *haikai* and tanka, beginning with his revisionism on Bashō. In "Furuike no gin" 古池の吟 (The meaning of the old pond, 1889), Shiki explains that "when reading Spencer's *Philosophy of Style* this past spring, I came upon a representation of the whole *image*, in other words, representing the whole with the part."[17] Spencer provided Shiki with the definition of synecdoche amidst a new arsenal of literary and poetic concepts—pleonasm, metonymy, hyperbole, personification, and apostrophe—not addressed in the literary theories of Yano or Shōyō.

Spencer's definitions of language include his conviction that the mother tongue of "primitive" Anglo-Saxon evokes more concrete images in the minds of English readers than the abstractions of Latin. On this nationalistic point, at least, Shiki did not concur. Despite his nationalist leanings and love of the *Manyōshū* poets, Shiki never felt compelled by a Norinaga-like obsession with Yamato words or spirit. Quite to the contrary, he advocated the use of everyday Chinese words whenever they provided a more concrete image than abstract or obscure Japanese ones. In *Nanatabi utayomi ni atauru sho* 七たび歌よみに与ふる書 (Seventh letter to tanka poets, February 28, 1899), Shiki argues for the need to consider Chinese words: "If we took away *The Tale of Genji*, *The Pillow Book*, and other works with Chinese words, what kind of Japanese literature would be left?"[18] Using an example whose importance is obvious in the post-*rakugo* transcription literary landscape, Shiki explores the differences between the Chinese word *botan* for "peony," and its Japanese equivalent, *fukamigusa*.

16. Ibid., 36. See Matsui, *Shasei no henyō*, 142.
17. *SZ*, 1:95. See also Matsui, *Shasei no henyō*, 136–40 and Akio's *Shiki no kindai*, 82–83. Translation is my own.
18. *SZ*, 10:30. Translation is my own.

He argues that *botan* has a more immediate impact and should be used in haiku and tanka. Regardless of the nationalistic leanings that led him to work for Kuga Katsunan's newspaper *Nippon* and his abortive stint as a war correspondent in China, Shiki's linguistic innovations make no political statement about China versus Japan, nor does he prioritize one literary style over others—the *Utayomi ni atauru shō* 歌詠みに当たうる書 (Letters to tanka poets, 1898–99) collection is actually written in the epistolary *sōrōbun*, consistent with its framing as a series of missives directed against *tsukinami* poetics.

In *The Origins of Modern Japanese Literature*, Karatani holds that Shiki liberated Japanese poetics from rhythm. While this may be true in the narrow sense of looking beyond poetic language or even the restricted 5–7 meter of traditional Japanese poetry,[19] one could make a far more bold statement about blank verse, or *shintaishi* (literally, "new-style poetry"), which was entirely free from conventional versification such as meter, rhythm, seasonal tropes, and the like. Shōyō in fact compares the novel to blank verse in its liberation from the 5–7 prose-poetic versification style called Bakin-*chō* (Bakin meter) after *gesaku* writer Takizawa Bakin. Parenthetically, I might add that blank verse, which began in 1882 with the publication of *Shintaishi-shō* 新体詩抄 (Collection of new-style poetry, 1882)—the same year as Takusari's shorthand—was a preferred genre of poetry for aspiring writers such as Tōson and Doppo, who later took up literary sketching and began writing novels in the unified style. Shiki, too, tried his hand at blank verse, but it was not a site of radical innovation or critique for him.

Shiki's most iconoclastic move was rather to introduce the principles of statistics to proclaim the exhaustion of traditional Japanese poetry of tanka and *haikai*. In the short essay "Go to shōgi" 碁と将棋 (The game of *go* and Japanese chess) from 1889 Shiki protests that *go* is not given equal respect simply because it is played by the lower classes. The same is true, he maintains of the difference between tanka and *haikai*. At the point where the game of *go* must end decisively, *shōgi* can be played with

19.   Shiki called for plain language and Sinologisms to revitalize traditional Japanese poetics (*uta*). Shiki's liberation of rhythm, if that is the right expression, is amply demonstrated by regularly employing techniques such as *ji amari*, which adds one more syllable than is normally permitted. See "Ji amari no haiku," 字余りの俳句 in *Dassai shoya haiwa*, reprinted in *SZ*, 4:163–65.

substantially more moves. To make absolutely clear the basis for his ideas, Shiki inserts in English the two words "combination" and "permutation" glossed with the words *sakuretsu* and *junretsu*, respectively.[20] Already Shiki was working toward a mathematical reconception of haiku. These statistical notions, and the strategies for mapping out movement as on the grid of a board game, would be thoroughly reworked into his critique of traditional poetics three years later in "Haiku no zento" 俳句の前途 (The future of haiku) from *Dassai Shōku Haiwa* 獺祭書屋俳話 (Talks on haiku from the otter's den, 1892):

> A certain contemporary scholar conversant with mathematics has said: "It is evident from the theory of permutations that there is a numerical limit to the tanka and haiku of Japan, which are confined to a mere twenty of thirty syllables."[21] In other words, sooner or later, the tanka . . . and haiku will reach their limit. He says that even now it has reached the point where not a single new poem is possible. . . . Although one may place the blame on the many mediocre teachers and poets who have appeared in this age of decline, part of it must certainly be assigned to the intrinsically narrow confines of the tanka and haiku. You may ask, "If that is so, when will the end come for the haiku and tanka?" And I reply: "I can't, of course, predict the time of their total extinction, but speaking approximately, *I think the haiku has already played itself out. Even assuming that the end is yet to come, we can confidently expect it to come sometime during the Meiji period.* The tanka allows more syllables than the haiku and thus, from the mathematical standpoint, the number of tanka possible is far greater than that of haiku. However, only words of the classical language may be used in the tanka and since there are extremely few, the tanka is in fact even more limited than the haiku. I conclude, therefore, that the tanka has been practically played out prior to the Meiji period.[22]

Needless to say, until Shiki no one in the Japanese tradition had ever expressed a crisis in these terms for the simple reason that no one conceived

20. *SZ*, 10:52. See also the short piece "Go-ron" 碁論 (Theory of *go*) in *SZ*, 10:82–83.

21. The Chinese characters 錯列法 are glossed with their English equivalent phoneticized in kana, パーミュテーション.

22. Beichman, *Masaoka Shiki*, 35–36. See also *SZ*, 4:165–66 (emphasis in original). The translation is Beichman's.

of its poetics as finite or exhaustible. This partly comes from a basic difference in the notion of originality touched upon in the previous chapter: operating within conventional boundaries versus striking out on one's own to create anew in the manner that defines modernity. It is a truism that modernity continues to define itself as change in opposition to conventionality, with each successive "movement" (Romanticism, Naturalism, and so on) retrospectively defined at the point its own set of conventions become institutionalized and static.

In his assaults on the complacency of Japanese poetics, it is altogether fitting that Shiki applied statistics, the science that depends upon standardization and principles of equivalence to flatten disparate phenomena into compliance. If statistics represented the model of dead language that he most vehemently opposed, it also became one of his chief rhetorical tactics. By the same logic of breaking down haiku and tanka into so many constituent parts of syllables, epithets, and other units of measure, the avid baseball fan who could no longer play the game converted dead poets into a registry of stats. Against the consistency and originality of Buson's masterful *hokku*, Bashō's batting average came down to a measly one hit in ten.

As Mark Morris succinctly puts it, "Modern haiku is born the moment haikai becomes archival."[23] But the archive was already foreclosed by Shiki, who decomposed *haikai* in order to compose haiku. Shiki's positionality as the end of *haikai* communality and the dawn of the poet as individual also corresponded with the final stage in the historical shift from *haikai-no-renga* (Bashō) to *hokku* (Buson) to *haiku* (Shiki): "In a modern industrial age Art is not the pursuit of groups, and Beauty beckons one poet at a time, if it beckons at all. And so haiku would become an art of individuals and the study of haikai a sifting through broken runes."[24]

It is well known that "Futatabi uta yomi ni atauru sho" 再び歌詠みに 当たうる書 (Further letters to tanka poets, February 14, 1899), which begins with the famous attack line "Ki no Tsurayuki was a lousy poet and the *Kōkinshū* was rubbish," launched a series of critiques against the canon of *haikai* and tanka poetics maintained by *tsukinami* haiku

---

23. Morris, "Buson and Shiki, Part I," 385.
24. Ibid., 385.

associations.[25] Yet it was also a forum for Shiki to explore ideas about Japanese poetry as modern literature, and to articulate some of his views on realism, which he primarily expressed in the last few years of his life (1900–1902) through artistic and literary sketching. In the sixth weekly installment from February 24, 1899, Shiki issues a figurative call to arms for modern literature to receive domestic and foreign investments just as the military did.

> Just as one would not build a fortress for the foundation of literature with commonplace *waka*, and just as one would not wage war with bows and arrows, or swords and spears, [reforms] must be carried out in the Meiji era. Today we pay out huge sums to foreign countries to purchase warships and cannons, for there is no other way to fortify Japan. That being the case, I would also see us invest some small amount of money to continue importing foreign literary ideas to fortify the fortress of Japanese literature. Life for *waka*, too, demands smashing old thought and introducing new thought. Therefore in our vocabulary I intend to use wherever it is appropriate ornate, vernacular, Chinese, and Western languages.
>
> *Jūrai no waka o mote Nihon bungaku no kiso toshi jōheki to nasan to suru wa kyūshi kensō o mote tatakawan to suru to onaji koto ni te Meiji- jidai ni okonawaru beki koto ni te wa nai no sōrō. Kyō gunkan o aganai daipō o aganai kyogaku no kane o gaikoku ni dasu mo hishō Nihon-koku o katameru ni hoka narazu, sareba kinshō no kingaku ni te aganai ebeki gaikoku no bungaku shisō sukui wa zokuzoku yunyū shite Nihon bungaku no jōheki o katametaku aru sōrō. Sei wa waka ni tsukite mo kyū-shisō o hakai shite shin-shisō o chūmon suru no kangae ni te shitagattte yōgo wa gago, zokugo, kango yōgo hitusyō shidai mochiuru tsumori ni sōrō.[26]*

While there are certainly resonances with the kind of saber-rattling rhetoric utilized by Isawa in his advocacy for imperial linguistics, Shiki's nationalism was considerably more benign. He did not seek to conquer foreign territory, but to shake up the moribund, domestic state of Japanese letters.

25. *SZ*, 7:23. As its name indicates, it was the second installment. The first, which valorized the *Manyōshū*, was simply titled *Uta yomi ni atauru sho* and dates to February 12, 1899. Translation is my own.
26. Ibid., 7:37. Translation is my own.

In *Zukan Nihongo no kindaishi* 図鑑日本語の近代 (Illustrated modern history of Japanese, 1997), Kida Junichirō notes the confusion caused by the adoption of the solar calendar in 1872 for *haikai* poetic circles. The rearrangement threatened to disrupt the balance between nature and culture by moving the dates of holidays and obscuring other occasions such as moon viewing. Kida observes that while one might expect the reinvigoration of seasonal words, in keeping with the spirit of "Enlightened Civilization," there was surprisingly little in the way of new developments due to the social hierarchies and cultural conservatism of *haikai* schools. Shiki did not feel bound by outdated constraints. In "Seventh Letter to Tanka Poets" Shiki uses the example of the train to argue for the proper inclusion of modern machine technology into poetic imagery. He argues that modernity should not be divorced from natural landscapes but rather incorporated into the visual field as its minor image to produce vibrant compositions:

> When told to compose poetry on unusual topics, many people hold erroneous ideas about how to approach trains, railroads, and other so-called "machines of civilization." For many unrefined people, the machines of civilization are a difficult subject about which to compose poetry, but if they won't do it, the task must fall to people of taste to create suitable combinations. A poem with no additional combinations, such as "the wind blows over the rails," comes across as exceedingly desolate. Alongside the rails one might have flowering reeds, or chestnuts scattered or pampas grass undulating in the wake of the train's passing—combining the image of the train with any number of other things should make it more visually appealing.[27]

It might be argued that because *kanshi* poets had already been dealing with these topoi for several decades, Shiki was merely bringing tanka and haiku in line with a common poetic freedom.[28] Yet his innovations were anathema to the *tsukinami* poetic circles that remained willfully closed off from modernity and clung to the glories of an imagined past. For Shiki, who does not mention the Chinese poets, the point was never the novelty factor, but to achieve the best composition of concrete images.

27.   *SZ*, 7:42. Translation is my own.
28.   See, for instance, Fraleigh, "Wang Zhaojun's New Portrait," 94–106.

While Morris dubs Shiki a "reclassicist," who anointed Buson as a saint formed in his own image, it might be better to follow this line of thinking and call Shiki a *reclassifier*, who sifted through the ruins and runes of old rules to construct a modern poetics.

After the successful serialization of *Talks from the Otter's Den* and his withdrawal from Tokyo Imperial University, Shiki became the literary editor and contributor for Kuga Katsunan's *Nihon* newspaper in 1892. He also served as the editor for its short-lived offshoot, *Shō Nihon* 小日本 (Little Japan), which lasted about six months. It was at these newspapers that Shiki made many of his critical contacts in the literary, poetic, and artistic communities. Shiki met the impoverished young painter Nakamura Fusetsu in 1895 through the introduction of Asai Chū, who contributed to the newspaper. Fusetsu would prove an enormously powerful influence on Shiki, introducing him to the principles and compositional techniques of sketching.

Asai was Fusetsu's teacher at the Kōbu Bijutsu Gakkō (Technical Art School), established in 1876 by government officials not ostensibly to teach fine arts, but to offer practical instruction in draftsmanship for engineering, industry, and other needs of the modernizing nation. From 1876 to 1878, the Italian artist Antonio Fontanesi taught there, and introduced his students, Asai and Koyama Shotarō among them, to European-style drawing and painting techniques. Fontanesi's legacy was the principles of sketching, whose dissemination very closely paralleled the emergence of shorthand. By the mid-1890s, the techniques Fusetsu taught Shiki had already permeated Meiji visual discourse as the first generation of *yōga*, or Western-style painting.[29]

Matsui Takako calls attention to Fontanesi's translated writings as the earliest evidence of a scopic regime that disseminated an idea of focal depth akin to the camera's lens. Fontanesi instructed students to concentrate upon the smaller order of details in the center of the composition while allowing the margins to remain less distinct. She insists that Shiki's essay *Jyojibun* 叙事文 (Prose writing, 1901) was consistent with this

29. Asai and Koyama constituted the opposition to Fenollosa and Okakura Tenshin's Orientalist aesthetics and nihonga movement at the Tokyo Bijutsu Gakkō (Tokyo School of Fine Art) from 1887 to 1896. From 1896, the Tokyo School of Fine Art under the new management of Kuroda Seiki and Kume Keiichirō taught Western-style painting.

compositional philosophy, isolating the key phrase: "although sketching is nothing more than copying things just as they really are, from the beginning there must always be some degree of selection" (*shasei to iu wa jissai ari no mama ni utsusu ni sōi nakeredo, moto yori tashō no sessha-sentaku o yō su*).[30] In the unfinished essay "Shasei, shajitsu" 写生、写実 (Sketching, realism, 1899), meanwhile, Shiki points out that the term *shasei* was nothing new to Japan; in one form or another, sketching had been around for more than a hundred years since the Sumiyoshi Tosa School, the pioneering work of Shiba Kōkan, and others. What changed, of course, was the discursive outlook. Artistic and literary sketching became a critical moment in the reimaginings of the nation that coincided with the standardizing spatial and temporal measures in the late nineteenth century. The Meiji-era artist Yoshida Fujio, whose husband Yoshida Hiroshi was a student of Koyama Shotarō and an associate of Fusetsu's, would recall that during the 1890s, when her husband traveled into the countryside to sketch landscapes, he was frequently mistaken for a government land surveyor by the locals.[31] From the standpoint of art education and imperial ideology, both specimens of interloper were not that far from the mark. In the same way that surveyors trained at the Technical Art School mapped the nation, landscape sketching, including what Raymond Williams and Karatani independently call "people as landscape,"[32] provided a cartography of the national imaginary.

We might juxtapose these uses of sketching in the late 1890s with those of the British Impressionist artist Alfred East, who visited Japan from March to June 1889 at the behest of the London Fine Arts Society. In addition to drawing, painting, and amateur photography, East made ample use of prose to record his journey.[33] There is a fluid movement

30. Matsui, *Shasei no henyō*, 238. *SZ*, 14:247. Translation is my own.

31. Tanaka, "Shasei ryokō," 906.

32. See Williams's discussion of this trope in George Eliot's *Mill on the Floss* in *The Country and the City*, as well as Karatani, *Origins of Modern Japanese Literature*, 24–25.

33. As noted in his obituary from the *Times* (September 13, 1913), East studied at the Ecole des Beaux Arts in Paris and exhibited his paintings in the early 1880s in the Paris Salon and the Royal Academy in London. He held memberships first in the Société des Arts Français and then its rival organization, Société des Beaux Arts. He was given an honorable mention in the Exposition Universelle in 1890, and knighted in 1901. He was also made an honorary member of the Meiji Bijitsu Kai after his trip to Japan. During his stay in Japan, he was feted by various luminaries, including Basil

between these different recording techniques that effectively demonstrates their mutuality as complementary tools in the artist's kit. Shortly after arriving at the port of Nagasaki from China on the first leg of his voyage, East stops off at a photographer's shop to have film developed. There he finds the occasion for a sketch in plain language of the glorious sunset framed by the rectangular entranceway of the photographer's shop.

> We went on to a photographer's shop to get some photographs developed which one of my friends had taken in China. I made a sketch from the door of the photographer's. It was as charming as any I did in Japan. There was a river coming down past the house which was situated on the river bank overlooking the harbour. To the right were big hills which at sunset were superb in color, their pale sandstone going from all the shades of rose to gradations of amber.[34]

In this admirably suggestive mixing of photographic, painterly, and verbal descriptive modes, East blurs the boundaries between these different compositional strategies. They seem equally valid, almost interchangeable forms of artistic expression. East's travelogue marks possibilities that were just on the cusp of emergence in Japan in the 1880s. It did not, however, circulate widely in his lifetime, and despite his welcome reception by leading figures in the Meiji artistic community, his travelogue had a negligible impact on modern Japanese literature. It was Shiki who translated techniques for representing the objectivist lens of the camera or painter's eye into literary sketching.

Shiki's experiment with literary sketching in the late 1890s also came about only after his career took him through a circuitous newspaper route. When *Shō Nihon* folded, Shiki briefly served a stint in the Diet as a reporter during the Sino-Japanese War (1894–1895). The work did not appeal to him,[35] and he volunteered as a war correspondent in China. Unfortunately for Shiki, the war ended shortly before he arrived, and his

---

Hall Chamberlain, Toyama Masakazu, and Ernest Fenollosa, and invited to speak before Japan's young generation of Western-style artists.

34. Cortazzi, ed., *A British Artist in Meiji Japan*, 18–19.

35. Although it is to be expected that Shiki became acquainted with shorthand reporters and may even have picked up some of their techniques, there is no record of these encounters (to my knowledge) in his writings.

tuberculosis was severely aggravated by the cramped, unsanitary quarters. During his brief time in China, where he was accompanied by Nakamura Fusetsu, he met Mori Ōgai, stationed as an army doctor. Whatever the cross-fertilization of ideas over poetry and literature that may have occurred is unknown, but Shiki remarks in *Shōen no ki* 小園の記 (Record of the little garden, 1898) that the several bags of flower seeds Ōgai gave him failed to bloom.[36]

Shiki returned to recuperate at Sōseki's house in Matsuyama, where Sōseki worked as a high school teacher before being sent to England by the Ministry of Education to study English literature. Shiki learned artistic sketching and developed literary sketching in 1898 even as he continued his poetic work on haiku and tanka. In works such as *Meshi o matsu aida* 飯を待つ間 (Waiting to eat, 1898) and *Shōen* 小園 (The little garden, 1898), Shiki objectively described the contents of his garden as one would a visual composition. As his health deteriorated, the garden would become his only window onto the world beyond his bedside.

Sketching from Life

Dead squid	*Sumi haite*
With the ink it spit out—	*ika no shiniiru*
low tide	*shiohi kana*

—Masaoka Shiki (1895)[37]

Attacking the poems of others	*uta o shoshiri*
And cursing others—	*hito o noshiri*
If you see my words,	*fumi o miba*
Still flowing along	*nao nagarete*
Know that I am still alive!	*yo ni ari to omoe*

—Masaoka Shiki (posthumously published, 1904)[38]

In his last years of life, 1900–1902, the interchange between life and literature, blood and ink, intensified considerably as Shiki's apprehension

---

36. *SZ*, 12:239. See Beichman's translation in *Masaoka Shiki*, 113.
37. Watson, trans., *Masaoka Shiki: Selected Poems*, 35.
38. *SZ*, 6:273. Translation is my own.

of his mortality affected every aspect of his literary, critical, and poetic corpus, which he determinedly set against the atrophy and degeneration of his physical body. Shiki continued to work across multiple registers in haiku and tanka, sketching, and criticism, even beginning a study group on the *Manyōshū* in 1900. It is therefore hardly possible to define Shiki, as Karatani does, as a Naturalist. I find Karatani's characterization problematic, not least because Naturalism did not exist in Japan until after his death. He offers as evidence Shiki's unsentimental approach to describing his deteriorating condition in *Byōshō rokushaku* 病牀六尺 (My six-foot sickbed, 1902). Yet Shiki's iconoclasm defies straightforward assignment to any single literary movement. The very same rationale in "sickness as meaning" that Karatani uses to locate Shiki's scientific objectivism toward his own decomposing body can be used to demonstrate Shiki's obvious Romantic idealizations of tuberculosis. He took the pen-name Shiki 子規 (Cuckoo)[39] in 1889 after he was diagnosed with tuberculosis, and named his literary magazine *Hototogisu* (Nightingale), both of which resonate with the bird whose throat bleeds as it sings in a quintessential Romantic trope that binds breath and voice, the immortality of the poetic spirit and the mortality of the flesh. It taps into the same Romantic wellsprings, if not the mawkish sentimentality, as Tokutomi Rōka's *Hototogisu* 不如帰 (1889), whose heroine dies of tuberculosis. Janine Beichman goes a step further, arguing that Shiki was counting on Takahama Kyoshi to become his disciple and take over the literary movements he founded after his death. In a brief period when Kyoshi declared his inability to fulfill that role, Shiki remonstrated with Kyoshi in a letter over his fear that his literature would perish in his death.

> In this letter, Shiki used "my literature" as a synonym for "my life." He had accepted the knowledge that he would die young, but he retained his attachment to life in sublimated form, as a wish for the survival of his literature. . . . It was as though he had resolved to create, as a substitute for his "child" Kyoshi, a literary corpus that would remain after he died. The

---

39. In a manner not uncommon in premodern times, he went through several name changes over the course of his life. Born Masaoka Tokoronosuke, he was known in his childhood as Masaoka Noboru and later as Masaoka Tsunenori.

rupture with Kyoshi proved to be transitory, but Shiki's sense of despera-
tion and tremendous personal stake in his literature only deepened. The
end result—an impulse to achieve the total merging of literature and life—
pervades the prose and poetry of his last years.[40]

The intensity of Shiki's identification of an interrelationship between life
and writing and the slippages of one into the other only deepened in his
final days. Again, one finds the example in Shiki's pleas to the editors of
Nihon for *My Six-Foot Sickbed* to remain in print when they surreptitiously
removed it in a bid to convince Shiki to give up his post:

> Dear Chief Editor,
>
> *Byōshō rokushaku* is now my life. When I get up each morning the pain
> is so bad that I could die. In the midst of it, I open the newspaper, see
> *Byōshō rokushaku*, and partially revive. Oh my suffering this morning when
> I looked in the paper—there was no *Byōshō rokushaku*, and I burst into
> tears. It was more than I could bear.
>
> If I could possibly get you to print just a little (even half) of an entry,
> you would be saving my life.
>
> I am so badly off that I must ask such a selfish thing.
>
> Masaoka Shiki[41]

Not only serial writing, but also serial reading gave him the motivation
to live. But there was nothing Naturalistic about Shiki's views other than
the objectivity of the narrative voice and gaze inherited from shorthand.

Despite the rapid deterioration in his health, Shiki persisted in
composing poems and sketches, even when he was physically incapaci-
tated and could only dictate them to others. Kyoshi served as the amanuensis for Shiki's literary sketch "September 14," written days before his
death. His final act on his deathbed was to compose three haiku in his
own hand. But there were already numerous instances when Shiki
identified the intersections of life and literature, as in the following haiku
from 1896:

40. Beichman, *Masaoka Shiki*, 25.
41. Morris, "Buson and Shiki, Part II," 318.

kogatana ya	The little knife—
enpitsu o kezuri	sharpening pencils with it,
nashi o muku	peeling pears[42]

Life and literature converge in the knife that precedes the cutting word (*kireji* 切字), the implement that slices fruit and sharpens pencils. The pencil was Shiki's lifeline and path to literary immortality, while fruit and morphine were his only sources of pleasure. Shiki likewise drew and painted fruit, flowers, and vegetables in watercolor and wrote literary sketches about the world contained in his garden. Much as the younger and healthier Shiki decomposed the archive to revitalize haiku and tanka, in his final years of debilitation, he made still-life sketches on the ripeness of fruit and vegetables with the same rapt attention he gave to documenting the decomposition of his own rotting body. Shiki's prodigious love of fruit was memorialized in several of Sōseki's novels, including *I Am a Cat*. In *Sanshiro*, Shiki posthumously becomes a veritable saint, or perhaps Taoist sage, the savior of students crippled by bad teachers and the inspiration for reaching beyond one's limits.[43]

Elsewhere in his poetry there appear moments when the discursive fluids of blood and ink intermingle on the same page:

Swatting mosquitoes—	*ka o utte*
blood stains	*gunsho no ue ni*
on the war tales I'm reading[44]	*chi o in su*

---

42. Watson, trans., *Masaoka Shiki: Selected Poems*, 51; *SZ*, 2:597.

43. As noted in chapter 3, on the train from Kumamoto to Tokyo, Sanshiro encounters a stranger (Professor Hirota), who explains that Taoist sages love peaches, and the poet Shiki was fond of all manner of fruit. But then comes a warning about the perils of Lamarckian piggishness. If one is too greedy, there is the danger of having one's nose grow to reach what one's hands cannot. Hirota likewise treats the incredulous Sanshiro to a spurious story about Leonardo da Vinci making poisoned peaches by injecting arsenic into a fruit tree. Sanshiro would again encounter "Shiki," the cuckoo, scribbled in his classmate Yojirō's notebook, this time escaping to freedom from the confines of the very sort of classroom where Shiki had flunked out. See *Natsume Sōseki zenshū*, 287–88, 312.

44. Watson, trans., *Masaoka Shiki: Selected Poems*, 56; *SZ*, 2:492.

The poet locates himself in the text, a mock-heroic inclusion where the act of reading parallels the sacrifice of soldiers risking their lives for their country (the patriotic glory denied Shiki on his abortive trip to China as a war correspondent for *Nihon*). In *Take no sato uta* 竹の里の歌 (Songs from a bamboo village),[45] a tanka compilation from 1898 to 1902, Shiki devoted two sequences of tanka in 1900 to the installation of glass windows on the veranda that enabled him to look out onto the garden. Transparency was not only textual effect, but also a precondition of composition. Shiki even composed poems about his frustration when condensation obscured the view. While this does not bespeak a materialist conception that spilled over into his writings on the concrete image or the phonetic transparency of the unified style in literary sketching, it registers Shiki's keen awareness for the fragility of his life and the limited horizons of his gaze toward the end.

Let good fortune come to the person   *tokobushi mi*
who installed these glass doors   *fuseru ashinae*
for me,   *waga tame ni*
now and forever   *garasudo harishi*
a bedridden cripple   *hito yo sachi are*[46]

Shiki also made himself into the subject of transcriptive realism in a discursive loop that included not only haiku, tanka, and literary sketching, but also self-portraits in watercolors and photography. Asai Chū, Nakamura Fusetsu, and others participated in the dialogic composition of this photomontage of Shiki's physical decline and indomitable will to live. Matsui Takako's study of Shiki explores some of aspects of this process, such as the multiplication of portraits from photograph to sketch, as well as the different angles and poses.[47] These were not merely aesthetic decisions. As Shiki's spine degenerated from tuberculosis, his back became severely hunched, he lost the use of his legs, and he was eventually unable to stand. Not surprisingly, the photographs and sketches increasingly

45. Shiki originally published his tanka in *Shō Nippon* using the pseudonym Takeno Satobito (Bamboo Villager).
46. *SZ*, 6:267. Translation is my own.
47. Matsui, *Shasei no henyō*, 214–21.

9.1 Akaseki Sadakura's photograph of Masaoka Shiki, April 1901 (from *Masaoka shiki zenshū*, vol. 6, i). Photograph courtesy of Matsuyama Shiki Memorial Museum.

deemphasize the body and concentrate on his face. I want to call attention to a photograph taken by Akaseki Sadakura on April 5, 1901 (figure 9.1), which is more inclusive of the painful, tragic realities of Shiki's situation, and which he in turn used to reflect on the problem of what is included and excluded from view. He produced two responses to the photograph. In his own hand, Shiki added an inscription that accompanies the photograph.

> On April 5, 1901 Akaseki Sadakura took my picture. I am propping my upper body up on my right elbow above my pillow and half of my body is outside the futon. At the side of my pillow are the manuscripts for haiku, tanka and *My Illness*. Hanging from the pillar is a straw raincoat. In a white vase are the buds of cherry blossoms. Just above the cherry blossoms a Chinese-made fan can be seen. Right in front of my pillow is an inkstone from Kokubunji Temple.
>
> *Meiji sanjūnen shigatu itsuka Akaseki Sadakura-shi no satsuei suru tokoro/*
> *Yo wa migi no hiji o makura no ue ni taku shite hanshin o futon no soto ni*

dashi-iri/ Makura moto ni aru wa haikō kakō Wagabyō *genkō nari. Hashira ni kakareru mino/ hakubin ni iketaru wa sakura no tsubomi/ Sakura no ue ni sukoshi mieraru wa Shina-sei no uchiwa/ Makura no sugu mae ni aru wa Kokubunji ga no suzuri.*[48]

The inscription is a quintessential literary sketch, objectively taking in the dying man and his surroundings with equal degrees of dispassion. Despite his use of the first person, Shiki made himself over into a third-person narrator, working like the shorthand reporter to simply take down the actuality, rather than the essence of its representation.

A much different take is evident in the tanka he also composed on the occasion, which betrays the knowledge of what is included and excluded from view beyond the limits of an objective gaze.

A photograph taken	*yamu ware o*
of me in my illness—	*utsusu shashin*
the cherry blossoms	*ni toko no he no*
in the vase near my bed	*bin ni sashitaru*
ended up in the picture	*sakura utsurinu*[49]

The juxtaposition of his illness with the beauty and fragility of the cut cherry blossom branches recuperates the tension between life and literature, beauty and decay, as well as modernity and tradition that occupies so much of his oeuvre. It is a rare instance of Shiki composing a poem about photography, which he largely chose not to thematize, with the significant exception of a tanka sequence based on photographs of him before leaving on the trip to China that irreparably damaged his health. Shiki poured his life into poetry and criticism, setting in motion the modernist shake-out of traditional poetics. Beyond the frame of his death, literary sketching would transform the modern Japanese novel in ways that Shiki, the "failed novelist," could scarcely have imagined.

48. *SZ*, 6. Translation is my own.
49. *SZ*, 6:303. Translation is my own.

CHAPTER 10

Scratching Records with Sōseki's Cat

Feline Amanuensis

To write down every event that takes place during a period of twenty-four hours, and then to read that record would, I think, occupy at least another twenty-four hours. Even for a cat inspired by literary sketching, I must confess that to make a literal record of all that happened in a day and night would be a *tour de force* quite beyond my capacities. Therefore, however much my master's paradoxical words and eccentric acts may merit being sketched from life at length and in exhaustive detail, I regret that I have neither the talent nor the energy to report them to the readers. Regrettable as it is, it simply can't be helped. Even a cat needs rest.

—Natsume Sōseki, *I Am a Cat* (1905)[1]

In the century after Edison's invention of the phonograph, scratching vinyl records signaled nothing so much as an annoying disruption to smooth playback and was a reminder of the fragility of the medium on which that music was stored.[2] So it remained until the late 1970s when

---

1. Epigraph from Sōseki, *I Am a Cat*, 205. I have slightly modified Graeme Wilson and Ito Aiko's translation as noted in the chapter. In Japanese, see *Natsume Sōseki zenshū* (hereafter *NSZ*), 1:183.

2. In addition to the fact that Pitman popularized the term "phonography" some thirty years earlier, it is important to remember that the technical apparatus was preceded by experimental devices such as Scott's phonoautograph (1857), whose first recordings of sound waves on soot-blackened paper were rediscovered and successfully

disc jockeys[3] in New York City transformed record-scratching from accidental abrasion into an art form. Moving the record back and forth under the stylus to control the beat and tempo became a signature gesture of the new genre of hip-hop. MCs would rap over the mixed and cut-up records, thereby creating new audial soundscapes and collages. In that historical moment, which marked an epistemic as well as rhythmic break from the age of Edison, what had previously been considered noise was incorporated into music. Likewise, a machine for playback became a musical instrument that could be played live or be recorded in its own right. Since the late 1980s, even as analog gave way to digital, turn-tabling has remained a powerful expressive tool that reclaims from mass culture what had become the thoroughly naturalized, passive reception of recorded sound.

Keeping the contemporary nuances of record scratching in mind, I invoke the phrase as a means to better understanding Sōseki's method in *I Am a Cat*, where the cat who narrates the story also periodically enters or readjusts the frame to critique its contents. Although the cat is not actually working with vinyl records, but with various assemblages of language, media, and anthromorphic agents, his mode of participation and commentary deftly exposes the artifice behind the supposed transparent phoneticity of the unified style as well as the claims of transcriptive realism to capture life "just as it is" (*ari no mama*) that came to dominate the literary mainstream by the turn of the century. In his own way, Sōseki effectively deconstructed the field of modern Japanese literature even as he became one of its most recognizable authorities.

The stand-alone short fictional work Sōseki originally published in 1904 for Masaoka Shiki's journal *Hototogisu* with the intent of gently lampooning literary sketching and a good many other aspects of Meiji-era "Civilization and Enlightenment" was met with such enthusiasm that he developed the narrative over the next two years into a sprawling novel of ten serialized installments. It traverses nearly all of the grounds covered as media history in this book, from the riotous profusion of postal, print, and inscriptive technologies to the multiplicity of written and spoken,

---

reproduced in 2008. See, for instance, Perlman, "Physicists Convert First Known Sound Recording."

3. The term was coined by American radio commentator Walter Winchell in 1935.

classical and vulgar forms of language at play in Meiji society. Japan's equivalent to "The Age of Edison" is likewise amply represented. Electricity veritably crackles through the pages with references to telegraph, telephone, and the wireless telegraph—the last having been known in English and Japanese simply as "wireless" before "broadcasting" or "radio" came into popular usage.[4] This is to say nothing of the mind-reading abilities of the cat achieved by sitting on his master's stomach and divining his thoughts by an electric current that passes through his fur. It is but one of many instances in which modern writing technology is not only personified but also represented as the agent of historical change. My reading of these elements intends to expose how *I Am a Cat* captured through parody and other means an episteme that had coalesced from the proliferation of new media and standardizing measures, obsessions with modern language, literature, and art, and a blending of humanistic knowledge from Western humanities, Chinese classics, Japanese poetics, and late Edo-early Meiji popular culture. Beyond cataloging the comical situations that arise from an incomplete or sometimes incomprehensible modernity, *I Am a Cat* is also keenly attuned to the limits of its own composition. In fact, the feline interlocutor/amanuensis constantly disrupts the narrative flow with reflections upon the instrumentality of writing itself.

Much of the scholarly discourse on discourse in *I Am a Cat* has vacillated between the primacy of speech and writing, from James Fujii's claims in *Complicit Fictions* that the eponymous cat "intones" the text to Kono Kensuke's *Shomotsu no kindai* which emphasizes the fundamentally inscriptive aspects of the narrative. As the second section of this chapter will explore, what has been overlooked in the binary opposition between orality and écriture is the literal middle ground of the nose. For Sōseki, the nose serves as an illocutionary organ that lies between the centers of modern vision ("eye" am a cat) and speech ("I" am a cat), yet belongs to neither. Contrary to the noble functions of eyes and mouth in establishing good meaning and sense, the nose is portrayed as a comical mess, the leaky faucet that effectively spoofs oratorical and literary pretensions. Indeed, it exposes the very limits of literariness, be

---

4. The terms "radio" and "broadcasting" came into popular use in the 1910s. "Radio," a contraction from "radiotelegraphy," is cognate with the verb "to radiate," while "broadcasting" was originally an agricultural term to spread seeds over a wide area.

it in the unvarnished realism of literary sketching, the minimalist expressive power of haiku, or the philosophical flights of fancy indulged in by the appropriately surnamed Kushami-sensei (Master Sneeze)[5] who adopts the cat.

The starting point for reappraisal of *I Am a Cat* depends on how we grasp the positionality of the modern subject against, or through, standardized national linguistic and ethnic identity. In *Complicit Fictions*, Fujii argues that *I Am a Cat* resists an increasingly compulsory Japanese citizenship and subjecthood, in spite of its composition in the unified style that best came to represent modern subjectivity. While I am broadly sympathetic to this thesis, I disagree with his understanding of that narrative position and its media-historical referents. For Fujii, *I Am a Cat* derives its punch from the quintessentially oral genres that preceded the unified style. As he argues, "Sōseki's inclination to resist a single, standardized language in his early prose explains his foregrounding of spoken language, which, in [*I am a Cat*] appears not only as the cat's direct discourse but as *rakugo*, popular songs, epigrams and a host of other forms with strong traces of orality."[6]

That the speaking subject or agent of history makes himself heard in the unified style further confirms the phonocentrist bias that privileges speech over writing. Fujii elaborates on this presumption of primary orality and its impact on the narrative voice of the text:

> The monologic (single-voiced) tendencies characterizing the new expression of subjectivity (through genbun itchi) clearly troubled Sōseki, and it is no accident that his earliest major prose narrative *Kusa makura* and [*I am a Cat*] implicitly challenged the standardization of such narratives. Years later when genbun itchi has become common practice, Sōseki's tale served to defamiliarize it by employing an orational style—a hybrid language inscribed with conventions of orality and scripted form—intoned by a household cat.[7]

---

5.   Given the centrality of the name "Sneeze" to the disquisition of noses, mouths, and what comes out of them, I break with the usual convention of keeping proper names intact, and use the English translation throughout my analysis. For consistency's sake, I likewise follow Aiko Ito and Graeme Wilson's translation of the text for other characters after indicating the original names in Japanese.

6.   Fujii, *Complicit Fictions*, 112.

7.   Ibid., 111.

While there is no disputing that Sōseki may have thematized forms of resistance to the standardization of the unified style, it is also true that he helped naturalize it in his best-selling novels *I Am a Cat* and *Bōtchan* (1906), as well as in other early works. Nor did he attempt to challenge its hegemony after its dominance as a *literary* style had become a historical fait accompli. Irrespective, the fundamental organization of the text is a deconstruction through writing rather than speaking.

Orality and oratory are fielded through the medium of the record; hybrid though it may be, the language is not intoned as the cat's meow, but taken in by the cat's scan and taken down by the cat's paw. Moreover, *I Am a Cat* is consistent with the principles of the classic readerly novel as defined by Roland Barthes in *S/Z*—namely, that the novel and its literary language hold the master keys that unlock all the codes of representation in the arts and sciences. Accordingly, it is not only the foibles of humans that come under scrutiny in *I Am a Cat*, but the humanities and the production of all humanistic knowledge. In this deceptively complex and polyphonic text, we can see how the artifice of a new literary language and other codes of representation were themselves objectified by a narrator who is also by turns a scribe and author in his own right.

The fact that the cat is not an orator in no sense negates the oratorical and aural precision he mobilizes as an amanuensis. Indeed, his transcriptive work powerfully recalls the many shorthand reporters at the turn of the century who midwifed the unified style primarily through their transcription of *rakugo* and *kōdan*. Along with his many asides and digressions that effectively disrupt the illusion of immediacy and point to the artifice of an omniscient and omnipresent narrator, the cat employs several strategies for representing speech: spelling out words such as difficult foreign names with dashes between the letters, as in A-n-d-r-e-a d-e-l S-a-r-t-o; instances of s-s-stammering; dialectical variations the neighborhood cats share with their humans, and so on. I should hasten to add that some of the less scrupulous reproductions of the text simply eliminate these dots, dashes, and orthographic irregularities as so many oddities. It is all the more reason we must return to the "original," or at least first editions of Meiji texts, as they were produced during that period of intense intellectual and material fermentation. I would maintain these dots and dashes are not merely technical effects, but telegraph

Sōseki's very deliberate project of deconstructing the narrative tendencies of the unified style.

Leaving aside the horribly tacky titles and innumerable spinoffs such as *I Am a Cat, Too,* and *I Am a Dog* that so offended the sensibilities of the sexually frigid narrator-diarist of Mori Ōgai's ヰタ・セクスアリス (*Vita Sexualis*, 1907),[8] let us put it as plainly as possible: *I Am a Cat* is written by a cat. Taking as its point of departure the observational mode of literary sketching and shorthand reporting before that, the text includes numerous instances that do not merely allude to, but in fact explicitly spell out in words and image, the fact that the cat is a scribe and/or writing machine. The frontispiece to the first edition of the text—that is, the version commissioned by Sōseki—is an ink drawing of a cat-headed Egyptian god holding in its hand a writing brush and an open book—the cat-god as scribe (figure 10.1).

This Art Nouveau–style illustration is made to resemble a hieroglyphic inscription with iconic representations of fish providing sustenance in the afterlife, and the title of the novel in faux woodcut style above the cat's head is consistent with the popularity of Egyptology and the debates over hieroglyphy previously noted. Writing provides an immortality that is preservation beyond the grave; if it is not already, as Kittler contends,[9] the language of the dead. Regardless of whether Sōseki explicitly intended to draw a parallel to the ancient Egyptian belief in the cat as a repository for the soul, the frontispiece lends further credence to the representation of the cat as an agent of writing. Likewise, Kono calls attention to the cover art and other illustrations in the text. The first edition of the text featured a giant cat-headed figure holding a long brush and puppet or chess-piece figures, presumably representing the characters in the novel.[10] Once again the valence of power is reversed, and the feline narrator becomes the godlike Creator, or puppet-master, of all that he surveys and writes down.

---

8. See also Yokota, "Wagahai-tachi mo 'Wagahai' de aru," on the ever-expanding catalog of horrible takeoffs (*Wagahai wa film de aru, Wagahai wa frockcoat de aru*, etc.) in *NSZ*, 1:8–12, supplementary insert.

9. See Kittler, *Gramophone, Film, Typewriter*, 8.

10. Kono, *Shomotsu no kindai*, 104.

10.1 Hachiguchi Goyō's illustration from the first edition of *Wagahai wa neko de aru,* 1905. Photograph courtesy of National Diet Library.

Further exemplifying the trope of cat-as-writer is a picture postcard delivered to Master Sneeze on New Year's Day. The postcard depicts a half-dozen or so cats engaged in lively acts of reading, writing, and dancing.

It is a printed picture of a line of four or five European cats all engaged in study, holding pens or reading books. One has broken away from the line to perform a Western dance singing *"it's a cat, it's a cat"* at the corner of their common desk. Above this picture, "I am a cat" is written thickly in Japanese black ink. And down the right side there is even a haiku stating, "on spring days cats read books and dance."[11]

11.   Sōseki, *I Am a Cat,* 25. *NSZ,* 1:24. I have modified the English translation to reflect Andō and Takamori's notes in the *zenshū* that the song "neko jya, neko jya" was popular in Edo since around the turn of the nineteenth century. The word "cat" was

Needless to say, the scene brings to a head many of the medial processes and scenes of writing raised earlier. The postcard itself attests to the rise of a new postal medium after the Russo-Japanese War (1894–1895), when the privately made postcard became both a popular art form and a tiny space for literary expression.[12] The cat is a sign in the Chinese zodiac, which marks one "premodern" mode of time-keeping and cosmology, but the cats on the postcard are in fact from "overseas" (*hakurai no neko*). It is worth noting that at least one cat is described as holding a pen, marking it as a writer. Akin to the unlikely coupling of the premodern *wagahai* with the *de aru* copula of the unified style, the juxtaposition of these disparate elements give the lie to linear notions of linguistic and cultural progress espoused by modernization theory.

It is impossible to sustain an argument that the cat is merely an orator when he constantly foregrounds his activities as a writer, complaining about how tiresome and difficult it is to keep track of everything for the reader. This is not a lament about the inadequacies of writing in general (once again, language representing the world), but the physical and psychological pressures imposed by literary sketching only one step removed from shorthand reportage. The amanuensis is literally a hand in the service of others:

> To write down non-stop every event that takes place during a period of twenty-four hours, and then to read that record would, I think, occupy at least another twenty-four hours. *Even for a cat inspired by literary sketching* [*ikura shaseibun o kosui suru wagahai demo*], I must confess that to make a literal record of all that happened in a day and a night would be a tour de force quite beyond the capacities of a cat. Therefore, however much my master's paradoxical words and eccentric acts may merit being sketched at length and in exhaustive detail, I regret that I have neither the talent nor

---

slang for a geisha, and the words to the song went something like "It's a cat, it's a cat, but can it be she's wearing geta, swaying along and wearing a plain-dyed robe?" (*neko jya, neko jya, neko ga geta haite, tsue tsuite, shibori no yukata de kuru mono ka*).

12. For instance, the journal *Hagaki bungaku* 葉書文学 (Postcard literature) was devoted to the postcard as a poetic genre, its use of haiku dovetailing with Edo-period practices of *surimono*, whose production and exchange circulated under somewhat similar circumstances.

the energy to report them [*hōchi*] to my readers. Regrettable as it is, it simply can't be helped. Even a cat needs rest.[13]

The pivot in his argument is a word play on the genesis of the text itself. On the one hand, the cat is claiming that he is a practitioner of literary sketching who has simply reached his productive limit. On the other, Sōseki debuted the first installment of *I Am a Cat* as a public reading (*rōdoku*) at the Sankai, a group associated with Shiki's journal *Hototogisu*. Hence the "cat" weaves in and out of his own writing as author and text, which reflects like a funhouse mirror on the conditions of its discursive production.

While Sōseki may have deviated in *I Am a Cat* from the conventional narratological order expected of the modern novel—a subject he had studied in England, taught as a professor of English at Tokyo Imperial University, and later published as *Bungaku Hyōron* (A critical study of literature, 1909)—he did so knowingly, with eye an toward the eighteenth-century origins of the genre by Laurence Sterne. Frequent references to *Tristram Shandy* that are both overt and intertextual weave through *I Am a Cat*, but arguably the most significant legacy of Sterne is his own knack for mastering, and then scratching, the dual effects of phonetic transparency and transcriptive realism.

In the essay "The Sentimental Journey" (1932) on Sterne's eponymous novel, Virginia Woolf remarks on his inimitably digressive and seemingly off-the-cuff style. It struck her as an almost literal transcription of speech with all the attendant concepts of immediacy and "true to life" recording that we have come to associate with modern Japanese literature:

> The jerky, disconnected sentences are as rapid and it would seem as little under control as the phrases that fall from the lips of a brilliant talker. The very punctuation is that of speech, not writing, and brings the sounds and associations of the speaking voice in with it. The order of the

13. Sōseki, *I Am a Cat*, 205. *NSZ*, 1:183 (my emphasis). I have modified Ito and Wilson's translation of the phrase from "Although I am all in favor of realistically descriptive literature" to "even for a cat inspired by literary sketching." In keeping with the original language, I have also changed their translation of *dokusha ni hōchi suru* as "set them all down for my readers" to the more straightforward "report them to my readers."

ideas, their suddenness and irrelevancy, is more true to life than to lit-erature. There is a privacy in this intercourse which allows things to slip out unreproved that would have been in doubtful taste had they been spoken in public. Under the influence of this extraordinary style the book becomes semi-transparent. The usual ceremonies and conventions which keep reader and writer at arm's length disappear. We are as close to life as we can be.

That Sterne achieved this illusion only by the use of extreme art and ex-traordinary pains is obvious without going to his manuscript to prove it. For though the writer is always haunted by the belief that somehow it must be possible to brush aside the ceremonies and conventions of writing and to speak to the reader as directly as by word of mouth, anyone who has tried the experiment has either been struck dumb by the difficulty, or way-laid into disorder and diffusity unutterable.[14]

Given the extensive relations already drawn out between Anglophone and Japanese media history, language, and script reform, by now it should not be altogether surprising to find that the rhetoric of transparency familiar to the unified style was already present in the English novel's most disrepu-table founding author. Likewise, Sterne mobilized illustrations, technical effects that disrupt the homogeneous surface of typography, supplementary prefaces, and afterwords that further confuse narrative good order. In his own name that needed no pseudonym, Sterne fashioned his own Roman-tic self-identification. Like Shiki, Sterne suffered from tuberculosis and chose as its symbol in the novel a small, singing bird—the starling, whose name in German is *stern*—that sings for its life and desperately tries to break free from the iron cage it is trapped in. Romantic breath that animates the spirit and invigorates both voice and body remains the central conceit of sentimentality.

Emblematic of the text's exposition and breakdown of these overlap-ping registers is the protagonist Yorick, a minor character from *The Life and Opinions of Tristram Shandy* (and, of course, *Hamlet*), who, true to his namesake, dies in the earlier text and returns in the sequel. In *Tris-tram Shandy* he is memorialized by the epitaph ⎡Alas, poor YORICK!⎤ on his gravestone. The words uttered with a sigh by every passerby in the

14. Woolf, *Collected Essays*, 96.

churchyard pile up and spill over onto the next page as a single, unremitting sheet of black. Indeed, the abundance of ink and pathos bleeds all the way through to the other side. Black on both sides is the condition of mourning where the excess of words and blank slate of their absence amounts to the same thing—abstraction in the impossible face of grievous loss.

Written after *Tristram Shandy* but chronologically set before it, *A Sentimental Journey* begins with Yorick disembarking at Calais for his journeys through France and Italy. Fresh off the boat, he sets down a characteristically rambling preface on the joys and dangers of "sentimental commerce," which also becomes a disquisition on the ills of solipsistic writing as being akin to masturbation. Yorick defines different types of travelers, above all the sentimental traveler, and what they must do to obtain the experiences they crave: pay at a premium what they could obtain more cheaply at home. Before the advent of the modern postal system that standardizes distances and rates and settles international affairs under the authority of the Universal Postal Union, the circulation and exchange of feeling are best obtained through sentimental commerce, which is possible only through travel. As Yorick remarks, "Knowledge and improvements are to be got by sailing and *posting* for that purpose."[15] However, in his haste to get things under way, Yorick composes the preface inside a *desobligeant*, a one-occupant carriage. He complains about the "Novelty of my vehicle," referring to the rickety and now steadily rocking carriage, but also to the novel as a means of disseminating his message. The episode reaches its climax when a pair of Englishmen pokes their noses into his chaise to see what the commotion is all about.

> We were wondering, said one of them, who I found, was an *inquisitive traveler*, what could occasion its motion.—'Twas the agitation, said I coolly, of writing a preface.—I never heard, said the other, who was a *simple traveler*, of a preface written in a *Desobligeant*.—It would have been better, said I, in a *Vis a Vis*.—*As an Englishman does not travel abroad to see Englishmen*, I retired to my room.[16]

---

15.   Sterne, *A Sentimental Journey*, 36 (my emphasis).
16.   Ibid., 37 (emphasis in original). The *vis-à-vis* is a carriage in which the occupants face one another.

The narrative that proceeds from this herky-jerky beginning takes license with all of the conventional appurtenances of the novel: linear unfolding space and chronological sequence, the pact of confidence and confidentiality between narrator and reader, and the conformity of language to a recognizable, mimetic relation to the world. For Sterne the solipsism and artificiality of the narrative voice is not allowed to pass unnoticed as intellectual masturbation. One cannot travel or write or be in this world alone. In advance of Sterne's latter-day rediscovery by poststructuralism, he served for Sōseki as an exemplar for decomposing and rearranging the narratological structures of the "transparently" realist novel.

The Discourse of Noses

In her study of manual and mechanical technologies of writing in the nineteenth century, Lisa Gitelman makes an incisive observation into the binary oppositions of physical and social organs that can also be extended to Soseki's *I Am a Cat*: "The negotiating table was set with powerful assumptions, among them the dichotomous oppositions of ear and eye, mouth and page, private and public, experience and evidence, 'man' and machine. These dualisms operated in tandem and in opposition in the appropriation of a sense of textuality somehow suited to the modern moment."[17] *I Am a Cat*'s discourse of the dichotomies of vision and speech, and what splits the difference between them, begins in the first pages of the text, immediately following the cat's now-famous pronouncement: "I am a cat. As yet I have no name." The cat's first encounter with a human being is the species called "student" (*shosei*). The comic disquisition on the nose, which Sōseki also triangulates off Sterne's *Tristram Shandy*,[18] depicts the students as having strange, bald faces with tiny slits from which small puffs of smoke occasionally emanate. The cat's defamiliarization of human beings has long been recognized as following in the tradition of Swift's *Gulliver's Travels* (1726), another founding text of the English

17. Gitelman, *Scripts, Grooves, and Writing Machines*, 14–15.
18. In *Tristram Shandy*, the nose is a signifier of sexual proportions, betokening intercourse, not the stickiness of what falls between oral and visual discourse.

canon that introduces faux taxonomies and cartographies, including of an imagined premodern Japan. In keeping with the fear of cannibals that pervades eighteenth-century European travel narratives, the feline narrator horrifically relays to the reader that the barbaric students are reputed to boil and eat cats. Adding to the humor, of course, was the fact that this was a time of dietary reform and meat-eating regimens. In any event, the cat's first memory of his life begins with a primordial cry in the dark, and the next thing he remembers he is resting in the palm of a student's hand, not unlike a pen or brush. It is an appropriate place to begin his narrative from the standpoint of an unnamed subspecies of cat we might call *felis domesticus amanuensis*.

Sometime later the cat finds himself literally brought into the picture of Master Sneeze's literary and artistic productions after he finds his way into the teacher's household. Part of the original installment that appeared in *Hototogisu*, this episode takes self-reflexive jabs at the practices of artistic and literary sketching. When the impressionable Sneeze is told in almost the exact words describing shorthand and sketching by one of his students that the sixteenth-century Italian master draftsman and painter Andrea del Sarto once remarked, "if you want to paint a picture, always depict nature as it is (*shizen sono mono o utsuse*),"[19] he decides to try his hand at watercolors to sketch the cat. The result is predictably more than a trifle ambiguous. The cat observes,

> I confess that, considering cats as works of art, I'm far from a collector's item. . . . But however ugly I may be, there's no conceivable resemblance between myself and that queer thing which my master is creating. . . . Furthermore, and very oddly, my face lacks eyes. The lack might be excused on the grounds that the sketch [*shasei*] is of a sleeping cat but, all the same, it is not at all clear whether the sketch is of a sleeping cat or a blind cat.[20]

In other words, Sneeze's drippy attempts land in the gray area between the blind objectivity of sketching and mere blindness. If the references to sketching were not obvious enough, the cat calls attention to Sneeze's attempts at cultivation in humanistic learning from contributions to

19.   Sōseki, *I Am a Cat*, 10. *NSZ*, 1:10.
20.   Sōseki, *I Am a Cat*, 10–11. *NSZ*, 1:10–11.

prominent Meiji literary magazines to classical Confucian and Edo lite-rati accomplishments.

> He can't restrain himself from trying his hand at everything and anything. He's always writing haiku and submitting them to *Hototogisu*; he sends off new-style poetry to *Meisei*; he has a shot at English prose peppered with gross mistakes; he develops a passion for archery; he takes lessons in chant-ing Noh play-texts; and sometimes he devotes himself to making hideous noises with a violin.[21]

Sneeze's failed accomplishments are an accumulation of genres and print media from Shiki and Sōseki's circle: haiku and literary sketching, oratorical arts (*wajutsu* or *wagei*), and the arrhythmic twanging of the archery and violin bows.

In his name and in his butchering of the noh chant and the violin, Sneeze powerfully evokes Fenollosa's remarks in "The Truth of Art" that making the strings of the koto groan and cry may be truthful, but it is hardly beautiful. The name "Sneeze" likewise reflects the bumbling nose and moustache that find their way into the scene of writing. For that matter, the feline narrator also finds repugnant the fact that Sneeze's wife sleeps with her mouth open. He describes the proper divisions of labor between the elocutionary and illocutionary organs, and the dan-gers, both aesthetic and sanitary, in confusing them:

> The mouth and the nose have their separate functions: the former is pro-vided for the making of sounds and the latter for respiratory purposes. However, in northern lands the human creature has grown slothful and opens its mouth as little as possible. One obvious result of this muscular parsimony is that northern style of tight-lipped speech in which the words would seem to be enunciated through the nostrils. That is bad, but it's even worse when the nose is kept closed and the mouth assumes the respiratory function. The result is not only unsightly, but could indeed, when rat shit drops from the rafters, involve real risk to health.[22]

---

21. Sōseki, *I Am a Cat*, 8–9. *NSZ*, 1:9.
22. Sōseki, *I Am a Cat*, 208. *NSZ*, 1:185.

These are only several of the instances where the nose sets off a digression into all manner of nonsensical ideas. There is, for instance, Kangetsu's (Coldmoon's) rambling riff in the manner of a "public lecturer" (*kōshakushi*) on aesthetics, Lamarckian evolution, and the significance of the nose to world history.[23] It also appears in the finale to a sequence of hilarious mock ditties that lambastes the "Yamato spirit" for the dubious ideology that it is by equating it with a *tengu*, a race of fabled long-nosed goblins:

There's not a person who hasn't said it	*Dare mo kuchi ni senu mono wa nai ga,*
There's not a person who hasn't seen it	*Dare mo mita mono wa nai.*
There's not a person who hasn't heard it	*Dare mo kita koto wa aru ga,*
But no one's ever come across it	*Dare mo atta mono ga nai.*
Yamato spirit—	*Yamato-damashi wa*
Is it just another variety of *tengu* out there?	*Tengu no tagui ka?*[24]

Returning to the critique of Mrs. Sneeze, the feline narrator continues his critique of human perspective owing to the literal narrowness of their vision: "Consider the human eyes. They are embedded in pairs within a flat surface and their owners, therefore, cannot simultaneously see to both their left and their right. It is regrettable, but only one side of any object can, at any one time, enter their field of vision."[25] The consequence of this blinkering, the cat insists, is that human beings are incapable of producing two identical copies of the same thing. It is the same with language. They learn by imitation from their mothers, nurses, and so forth, yet language shifts over time. Polysemy abounds, and the only constant is change.

From the acquisition of language and visual perception, the narrative segues to distinguishing sense from nonsense. The drivel that comes out of mouths, especially from supposedly learned individuals, is especially singled out for ridicule. Rather than claim the unified style as a source of constancy, the text employs disinformation as a strategy

23. Sōseki, *I Am a Cat*, 148–55. *NSZ*, 1:131–36.
24. *NSZ*, 1:131–36. Translation is my own.
25. Sōseki, *I Am a Cat*, 215. *NSZ*, 1:191.

of resistance against standardization and passive homogeneity. Meitei (Waverhouse), the mischievous graduate student and trickster, passes on to an undergraduate the spurious story that Nicholas Nickleby advised Gibbon to write his magnum opus *The History of the French Revolution* in English rather than French. The episode at once calls attention to the competition of national languages and blind recording loop of the Republic of Scholars: "Now this undergraduate was a man of almost eidetic memory, and it was especially amusing to hear him repeating what I told him, word for word and in all seriousness, to the oratorical assembly of the Japan Literary Society [*Nihon bungakkai no enzetsukai*]. And d'you know, there were nearly a hundred in his audience, and all of them sat listening to his drivel with great enthusiasm!"[26]

What begins as a harmless prank is thus recorded and broadcast through the mouth of the unwitting undergraduate, legislated from falsehood into fact, and exponentially broadcast at no less eminent a body than the Japan Literary Society. It was an acerbic take on the transcriptive disseminations via oratory not only of dispassionately observed realities but outright fabrications. The adversarial nature of Waverhouse to unguarded scholars should be a warning to those who trust without empiric verification regarding even the simplest bits of trivial knowledge.

In a related vein, Waverhouse ridicules the pomposity of know-it-all-professors, including his favorite target, Sneeze himself.[27] Set in motion by his vociferous attacks against Mrs. Sneeze's nose and his frequent references to *Tristram Shandy*, Waverhouse argues over Coldmoon's marital prospects, denouncing the wealthy man Kanehara (Goldfield) who would marry his daughter off to Coldmoon after he finishes his thesis simply for the prestige of the doctorate. Consistent with the rambling and deliberately nonlinear style characteristic of the text, Waverhouse maintains that Coldmoon's successful completion of his thesis will not only

26. Sōseki, *I Am a Cat*, 19. *NSZ*, 1:19.

27. Sneeze, who has already been nicknamed the "Maestro-of-the-Watercloset" for practicing noh chants in the bathroom, comes in for more toilet humor by Waverhouse, who suggests that meditating on the water stains from the leaky roof of the outhouse reveals a great natural masterpiece worthy of being sketched: "In a lavatory, for instance, if absorbedly one studies the rain leaks on the wall, a staggering design, a natural creation, invariably emerges. You should keep your eyes open and try drawing from nature. I'm sure you could make something interesting" (*I Am a Cat*, 20; *NSZ*, 1:20).

make such a financial union superfluous, but also generate a literary Big Bang whose reverberations will be felt throughout academia:

> Goldfield's merely a paper man, a bill of exchange with eyes and a nose scrawled onto them. If I may put it epigrammatically, the man's no more than an animated bank-note [*katsudō heishi*]. And if he's money in motion, currency one might say, his daughter's nothing but a promissory note. In contrast now, let us consider Coldmoon. . . . If I may adapt to Cold-moon's case one of my own earlier turns of phrase, I should describe him as a circulating library [*katsudō toshokan*]. He is a highly explosive shell, perhaps only a twenty-eight centimeter, but compactly charged with knowledge. And when at the properly chosen time this projectile makes it impact upon the world of learning, then, if it detonates, detonate it will.[28]

As these episodes reveal, the text is composed not solely as a linear narrative, but as a record intermittently scratched and spun by its narrator, the cat. While Sōseki obviously did not employ the hip-hop anachronism, he made ample use of electricity as another signifying process. Eyes like electric flashing signals, akin to railroad semaphores, electric beams that enable mental telepathy, and even Coldmoon's quip to name his dissertation "The Effects of Ultraviolet Rays on the Galvanic Action in the Eyeball of the Frog" round out the proliferation of writing technologies beyond ink on paper. Responding to readers' nitpicking questions toward the conclusion of the second volume about how Master Sneeze's thoughts and feelings could possibly be divined by the cat, Sōseki did not turn to spirit mediums, but modern media:

> I am a cat. Some of you may wonder how a mere cat can analyze his master's thoughts with the detailed acumen which I have just displayed. Such a feat is a mere nothing for a cat. Quite apart from the precision of my hearing and the complexity of my mind, I can also read thoughts. Don't ask me how I learned that skill. My methods are none of your business. The plain fact remains that when, apparently sleeping on a human lap, I gently rub my fur against his tummy, a beam of electricity is thereby

---

28. Sōseki, *I Am a Cat*, 199–200. *NSZ*, 1:178.

generated, and down that beam into my mind's eye every detail of his in-
nermost reflections is reflected.[29]

Reiterating the opening phrase of the novel, the passage provides cover-
age for discrepancies no doubt brought up by excessively literal readers
capable of suspending disbelief about a cat who can write and talk, but
who drew the line at mind-reading. The "beam of electricity" that settles
the matter is distinctly reminiscent of the invisible communication of
wireless telegraph, not to mention motion pictures projected by the earliest cinematic apparatus.

This play upon telepathic communication via electric or radio waves
resurfaces in a rambling conversation that brings Masaoka Shiki into the
picture. Waverhouse makes an offhand comment that his haiku are so
excellent that they left even the late Masaoka Shiki speechless. When
pressed if he actually knew Shiki, he coolly replies, "While we never actually met, we always communed by radio waves" (*nani awanakutte mo shijū
musen denshin de kantan aiterashite ita mon da*).[30] The wireless telegraph,
or radiotelegraphy, is effectively recast as a mode of radio—not
spiritualist—telepathy.[31] Shiki is not the foil in this operation, but the reflecting dish for Waverhouse's egotism. More salient is the figure of modern media once again enabling invisible or disembodied communication.

Lastly, I want to turn to the scene of composition at the core of the
novel that decenters writing. The cat observes Sneeze trying in his usual
half-baked fashion to create something profound. Sneeze sets out to write
an epitaph for Sorosaki, another alter ego for Sōseki, who is described as
having died of peritonitis while undertaking a postgraduate study of the
theory of infinity. Sneeze's attempts at philosophical poetry set in motion
a series of entanglements. His undisciplined mind exhausted by the effort,
he shifts from creative inspiration to critical exasperation and crosses out
his own words. Words that begin with the wetting of a brush from the lips
beget writing that soon digress into a drawing of a mouth, and returns the
brush to the mouth. Even his whiskers are roped into the act of clumsy

29.   Sōseki, *I Am a Cat*, 474. *NSZ*, 1: 406–7.

30.   *NSZ*, 1:513. Translation is my own. See also the wordplay on 477.

31.   Wilson and Ito incorrectly translate "wireless telegraph" as "a kind of spiritual
telepathy" (*I Am a Cat*, 579).

calligraphic writing. Failing to articulate anything poetic or profound, Sneeze shifts discursive gears into drawing and then mere scribbling:

> My master, brush in hand, racks his brains, but no bright notions seem to emerge for now he starts licking the head of his brush. I watched his lips acquire a curious inkiness. Then, underneath what he had just written, he drew a circle, put in two dots as eyes, added a nostrilled nose in the center, and finally drew a single sideways line for a mouth. One could not call such creations either *haiku* or prose.[32]

At a glance it is calligraphic literati art at its worst, lacking in the spirit or external media capture of anything true or beautiful. Ironically, Sneeze's incompetence as an amateur scholar-poet does produce a deep signifying gesture, albeit not one of which he is aware: he produces an unflattering self-portrait.

Sneeze's self-caricature brilliantly realizes the novel's recurring motif about the now-overlapping, now-separate nature of visual and verbal regimes in the Meiji period being separated only by a nasal drip. This smiley face is the culmination of the disquisition on the faculties of sight, sound, and what lies in-between. Since the nose is in the center of the face, it should be the important place or have the most important function. Instead it is a running joke.

The scene continues with an absurd attempt at transparent description in the unified style. When this, too, results in nonsensical prose, Sneeze takes the expedient step of drawing lines through the botched composition. Before it can approach the total blackout along the lines envisioned by Sterne, however, Sneeze's nose, or more precisely his nose hair, intervenes.

> After further moonings, he suddenly started writing briskly in the colloquial style [*genbun itchi-tai*]. "Mr. the-late-and-sainted-Natural Man [*tennen koji*] is one studies infinity, reads the *Analects of Confucius*, eats baked yams, and has a runny nose." A somewhat muddled phrase. He thereupon reads the phrase aloud in a declamatory manner and, quite unlike his usual self, laughed. "Ha-ha-ha. Interesting! But that 'runny nose' is a shade cruel,

32. Sōseki, *I Am a Cat*, 95. *NSZ*, 1:88.

so I'll cross it out," and he proceeds to draw lines across that phrase. Though a single line would clearly have sufficed, he draws two lines and then three lines. He goes on drawing more and more lines regardless of their crowding into the neighboring line of writing. When he has drawn eight such obliterations, he seems unable to think of anything to add to his opening outburst. So he takes to twirling his moustache, determined to wring some telling sentence from his whiskers.[33]

James Fujii cogently analyzes some of the multiple layerings of the generative act and scene of writing demonstrated in this episode. He is quick to attend to the interpenetrating regimes of visuality and orality that constantly leave a material or graphic remainder with each failed attempt at evoking poetic grandeur:

Here, where language production is event, we witness Kushami's writing degenerate into doodling. Kushami begins with an affected poetic conceit, but the resistance encountered in the act of writing redirects his efforts to an epitaph for the natural saint; mid-sentence his writing takes a comic turn. The lines striking out the comic description of the runny nose exceed their simple copy-editing function to become graphic artifacts in their own right—the eight parallel lines that stretch beyond the boundaries of the writing lines. Words here are not merely referential; they are themselves productive, participating in and altering events.[34]

Of course, words are never merely referential. Omnipresent in this scene of writing are excesses and offshoots of a signifying process that has gone off the rails—a smoke-billowing différance engine.

And the scene keeps chugging along. The mustache serves as a parody of the traditional brush: connected to the nose and wetted by the mouth, it only adds to the shortcomings of the would-be artist-poet. Even Sneeze's love of jam is included in the nasal humor, as his wife peevishly complains about his excessive consumption of the imported commodity. After repeatedly crossing out his nonsensical ruminations, Sneeze plucks several hairs from his nose and sets them down on the writing pad. Upon

33. Sōseki, *I Am a Cat*, 95. *NSZ*, 1:88–89.
34. Fujii, *Complicit Fictions*, 119.

discovering a pure white bristle, he pulls it out and shows it to his wife. This new bit of inspiration leads him to further edit his composition, blotting out all but the first sentence "—" and changing the rest into a clumsy *bunjinga*-style study of an orchid. Once again the scene recalls, and perhaps mocks, the bugbear in Fenollosa's "The Truth of Art." Turning over the ruined page, Sneeze at last completes a nonsensical epitaph for his friend Sorosaki, sagaciously summing it up thusly: "Born into infinity, studied infinity, and died into infinity. Mr. the-late-and-sainted Natural Man. Infinity."[35]

Sorosaki would appear to be another thinly veiled, self-deprecating reference to Sōseki, who suffered from stomach trouble, not to mention a supposed nervous breakdown, and died of a stomach ulcer about a decade later, in 1916. Biographical speculation aside, this figure, too, plays upon the endless digression of knowledge that slides across media, genres, and other signifying practices, but cannot be easily digested. It is Sneeze's final word on the epitaph worthy of inscription on a tomb—the most stable and lasting document-turned-monument. Instead, Waverhouse quips that it would be more appropriate to have it "engraved on a weight-stone for pickles and then leave it at the back of the main hall of some temple for the practice-benefit of passing weight lifters."[36] In other words, the words are of no value whatsoever, not even as portentous as the stone they are written upon.

Words mean nothing unless they are read and appreciated. The parodic reinvention of the amanuensis in the illustration of the cat-god as scribe makes its return in the third volume, where the feline narrator evinces a distinctly premodern sensibility that defends the plenary significance of his writing to lazy and disinterested serial readers. He ostensibly pays lip service to classical Chinese learning in response to the perceived iniquities of Meiji print culture, but it is in fact the last ditch appeal for the Romantic conceit of the author as the omnipotent Creator, whose compositions manifest and reveal the truth of the world.

Every single letter, every single word that I set down implies and reflects a cosmic philosophy and, as these letters and words cohere into sentences

35. Sōseki, *I Am a Cat*, 98. *NSZ*, 1:91.
36. Sōseki, *I Am a Cat*, 99. *NSZ*, 1:92.

and paragraphs, they become a coordinated whole, clear and consistent, with beginnings and ends skillfully designed to correspond and, by that correspondence, to provide an overall world-view of the condition of all creation. Thus, these close-written pages, which the more superficial minds amongst you have seen as nothing better than a tiresome spate of trivial chit-chat, shall suddenly reveal themselves as containing weighty wisdom, edifying homilies, guidance for you all. I would therefore be obliged if you would have the courtesy to sit up straight, stop lolling about like so many sloppy sacks, and, instead of skimming [*tsūdoku*] through my text, study it with close attention [*seidoku*]. May I remind you that Liu Tsung-yuan thought it proper to actually lave his hands with rose-water before touching the paper lucky enough to carry the prose of his fellow poet and fellow scholar, Han T'ni-chih. The prose which I have written deserves a treatment no less punctiliously respectful. You should not disgrace yourselves by reading it in some old dog-eared copy of a magazine filched or borrowed from a friend. Have at least the grace to buy a copy of the magazine with your own money. As I indicated at the beginning of this well-constructed paragraph, I am about to describe an aftermath. If you think an aftermath could not possibly be interesting and consequently propose to skip reading it, you will most bitterly regret your decision. You simply must read on to the end.[37]

The call to readers to sit up and act right invokes the specialized, but mutually opposing, styles of reading that came with the proliferation of new techniques and technologies of writing: skimming and close reading. The downside of scratching and scribbling was a corresponding loss of attention to the record as a carefully constructed totality.

37. Sōseki, *I Am a Cat*, 400–1. *NSZ*, 1:340–41.

Bibliography

*All are print materials unless otherwise noted.

Adas, Michael. *Machines as the Measure of Men: Science, Technology, and Ideologies of Western Dominance*. Ithaca, NY: Cornell University Press, 1989.

Akio, Bin. *Shiki no kindai*. Tokyo: Shinyōsha, 1999.

Albright, Robert. "The International Phonetic Alphabet: Its Backgrounds and Development," *International Journal of American Linguistics* 24, no. 1, pt. 3 (1958).

Anderson, Benedict. *Imagined Communities*. New York: Verso, 1991.

——. *The Spectre of Comparisons*. New York: Verso, 1998.

Anderson, Kenneth Mark. "The Foreign Relations of the Family State: The Empire of Ethics, Aesthetics and Evolution on Meiji Japan." Ph.D. diss., Cornell University, 1999.

Aoki, Shigeru, ed. *Meiji nihonga shiryō*. Tokyo: Chūō Kōron Bijutsu Shuppansha, 1991.

——, ed. *Meiji yōga shiryō: kaisōhen*. Toyko: Chūō Kōron Bijutsu Shuppansha, 1986.

Armitage, John. "From Discouse Networks to Cultural Mathematics: An Interview with Friedrich A. Kittler," *Theory, Culture & Society*, 23, nos. 7–8 (2006): 17–38.

Armstrong, Timothy. *Modernism, Technology and the Body*. Cambridge, UK: Cambridge University Press, 1998.

Aston, W. G. "Has Japanese an Affinity with Aryan Languages?" *Transactions of the Asiatic Society* 2 (1882): 199–206.

Azuma, Hiroki. *Sonzaironteki, yūbinteki: Jacques Derrida ni tsuite*. Tokyo: Shinchōsha, 1998.

Bailey, Richard W. *Images of English: A Cultural History of the Language*. Cambridge, UK: Cambridge University Press, 1992.

Baker, Alfred. *The Life of Sir Isaac Pitman*. London: Pitman & Sons Ltd., 1908.

Balibar, Etienne, and Immanuel Wallerstein. *Race, Nation, Class: Ambiguous Identities*. New York: Verso, 1991.

Beichman, Janine. *Masaoka Shiki*. Boston: Twayne Publishers, 1982.

Bell, Alexander Graham. "The Metric System: An Explanation of the Reasons Why the United States Should Abandon Its Heterogeneous Systems of Weights and Measures." Reprinted from *National Geographic* (March 1906). Washington, DC: Judd & Detweiler, 1906.

Bell, Alexander Melville. *Visible Speech: The Science of Universal Alphabetics*. London: Simpkin Marshall, 1867.

———. *World-English: The Universal Language*. New York: N. D. C. Hodges, 1888.

Benjamin, Walter. *Illuminations*. Translated by Harry Zohn. New York: Schocken Books, 1968.

Blaise, Clark. *Time Lord: Sir Sanford Fleming and the Creation of Standard Time*. New York: Pantheon Books, 2000.

Bluhm, Andreas, and Luise Lippincott, eds. *Light! The Industrial Age 1750–1900*. New York: Thames and Hudson, 2000.

Bowie, Theodore. *The Drawings of Hokusai*. Bloomington: Indiana University Press, 1964.

Bowring, Richard. *Mori Ōgai and the Modernization of Japanese Culture*. New York: Cambridge University Press, 1979.

Braisted, William, ed. and trans. *Meiroku Zasshi: Journal of Japanese Enlightenment*. Cambridge, MA: Harvard University Press, 1976.

Briggs, Asa, and Peter Burke. *A Social History of the Media: From Gutenberg to the Internet*, 3rd ed. Malden, MA: Polity, 2009.

Chamberlain, Basil Hall. *A Handbook of Colloquial Japanese*, 2nd ed. Tokyo: Hakubunsha, 1889.

———. *Practical Introduction to the Study of Japanese Writing*. London: S. Low, Marston & Co., 1899.

Charney, Leo, and Vanessa Schwartz, eds. *Cinema and the Invention of Modern Life*. Berkeley: University of California Press, 1995.

Cheyfitz, Eric. *The Poetics of Imperialism: Translation and Colonization from* The Tempest *to* Tarzan. Philadelphia: University of Pennsylvania Press, 1997.

Chisolm, Lawrence W. *Fenollosa: The Far East and American Culture*. New Haven, CT: Yale University Press, 1963.

Chōrin, Hakuen. *Nasake sekai mawari dōrō*. Tokyo: Dōmei Shobō, 1886.

Commager, H.S. "Schoolmaster to America." In *Noah Webster's American Spelling Book*, edited by H.S. Commager. New York: Teachers College Press, 1958.

Conrad, Joseph. *The Secret Agent: A Simple Tale*. New York and London: Harper and Brothers, 1907.

Copeland, Rebecca. *Lost Leaves: Women Writers of Meiji Japan*. Honolulu: University of Hawai'i Press, 2000.

Cortazzi, Hugh, ed. *A British Artist in Meiji Japan: Alfred East*. Brighton, UK: In Print, 1991.

Crary, Jonathan. *Techniques of the Observer: On Vision and Modernity in the Nineteenth Century*. Cambridge, MA: MIT Press, 1999.

Cross, Barbara. "Representing Performance in Japanese Fiction," *SOAS Literary Review AHRB Centre Special Issue* (Autumn 2004): 1–20.

De Bary, Theodore, Carol Gluck, and Arthur Tiedemann. *Sources of Japanese Tradition*, vol. 2. New York: Columbia University Press, 2005.

Debray, Régis. *Media Manifestos: On the Technological Transformation of Cultural Forms*. Translated by Eric Rauth. New York: Verso, 1996.

Derrida, Jacques. *Of Grammotology*. Translated by Gayatri Spivak. Baltimore: Johns Hopkins University Press, 1976.

———. *Papier Machine: Le ruban de machine à écrire et autres réponses*. Paris: Galilée, 2001.

———. *The Post Card: From Socrates to Freud and Beyond*. Translated by Alan Bass. Chicago: University of Chicago Press, 1987.

———. "Sending: On Representation," *Social Research* 49, no. 2 (Summer 1982): 294–326.

———. *Speech and Phenomena*. Translated by David Allison. Evanston, IL: Northwestern University Press, 1973.

———. "Speech and Writing According to Hegel." In *G.W. F. Hegel: Critical Assessments*, edited by Robert Stern, 2:455–77. New York: Routledge, 1993.

Dore, R. P. *Education in Tokugawa Japan*. Berkeley: University of California Press, 1965.

Doyle, Arthur Conan. *A Study in Scarlet*. London: Ward Lock & Co, 1887.

Duncan, David., ed. *Life and Times of Herbert Spencer*, 3 vols. New York: D. Appleton & Co., 1908.

Ellis, Alexander J, ed. *The Phonetic News*, January 6, 1849.

Ellmann, Richard, ed. *The Artist as Critic: Critical Writings of Oscar Wilde*. Chicago: University of Chicago Press, 1982.

Enchō, Sanyūtei. *Kaidan botan dōrō*. Tokyo: Tokyo Haishi Shuppansha, 1884.

Eppstein, Ury. *The Beginnings of Western Music in Meiji Era Japan*. Lewiston, NY: E. Mellen Press, 1994.

Ericson, Stephen J. *Sound of the Whistle: Railroads and the State in Meiji Japan*. Cambridge, MA: Harvard East Asian Monographs, 1996.

Faubton, James, ed. *Power: The Essential Works of Michel Foucault, 1954–1984*, vol. 3. New York: The New Press, 2001.

Fenollosa, Ernest. *Epochs of Chinese and Japanese Art*, 2 vols. New York: Dover, 1963.

———. *The Chinese Written Character as a Medium for Poetry*. New York: Arrow Editions, 1936.

Figal, Gerald. *Civilization and Monsters: Spirits of Modernity in Meiji Japan*. Durham, NC: Duke University Press, 1999.

Formanek, Susanne, and Sepp Linhart, eds. *Written Texts—Visual Texts: Woodblock-Printed Media in Early Modern Japan*. Amsterdam: Hotei Publishing, 2005.

Foucault, Michel. *The Archaeology of Knowledge and the Discourse on Language*. Translated by A. M. Sheridan Smith. New York: Pantheon Books, 1972.

———. *The Order of Things: An Archaeology of the Human Sciences*. New York: Vintage Books, 1994.

Fraleigh, Matthew. "Wang Zhaojun's New Portrait." In *Reading Material: The Production of Narrative Genres and Literary Identities* (PAJLS), vol. 7, edited by Dennis Washburn and James Dorsey, 94–106. West Lafayette, IN: Association for Japanese Literary Studies, 2007.

Frank, Manfred. *What Is Neostructuralism?* Translated by Sabine Wilke and Richard Gray. Minneapolis: University of Minnesota Press, 1989.

————. *The Philosophical Foundations of Early German Romanticism.* Translated by Elizabeth Millan-Zaibert. New York: State University of New York Press, 2004.

Fujii, James. *Complicit Fictions: The Subject in the Modern Japanese Prose Narrative.* Berkeley: University of California Press, 1993.

Fujikura, Akira. *Kotoba no Shashin o Tore.* Tokyo: Sakitama Shuppan-kai,1982.

Fujimoto, Yōko, ed. *Wakakibi no Nihon bijutsu: Meiji-ki no zuga kyōikusho to gakatachi.* Tokyo: Ibaraki-ken Kindai Bijutsukan, 1995.

Fujitani, Takashi. *Splendid Monarchy: Power and Pageantry in Modern Japan.* Berkeley: University of California Press, 1996.

Fukuoka, Maki. "Toward a Synthesized History of Photography: A Conceptual Genealogy of Shashin." *Positions:East Asia Cultures Critique* 18, no. 3 (2010): 571–97.

Galison, Peter. *Einstein's Clocks, Poincaré's Maps: Empires of Time.* New York: W.W. Norton, 2004.

Geijutsu Shinchō, ed. "Bakumatsu-itchi no kōkishin otoko: Shima Kakoku koko ni ari!" *Geijutsu Shinchō* 9 (1994).

Gerow, Aaron. *Visions of Japanese Modernity: Articulations of Cinema, Nation, and Spectatorship, 1895–1925.* Berkeley: University of California Press, 2010.

Gitelman, Lisa. *Scripts, Grooves, and Writing Machines: Representing Technology in the Edison Era.* Stanford, CA: Stanford University Press, 1999.

Gitelman, Lisa, and Geoffrey B. Pingree, eds. *New Media, 1740–1915.* Cambridge, MA: MIT Press, 2003.

Go, Komei. "Isawa Shūji to shiwahō." *Kyoto Seika University Research Laboratory Research Results & Publications* 26: 146–161. http://www.kyoto-seika.ac.jp/researchlab/wp/wp -content/uploads/kiyo/pdf-data/no26/go.pdf.

Goldstein, Sanford, and Seishi Shinoda. *Songs from a Bamboo Village: Selected Tanka from Takenosato Uta by Masaoka Shiki.* Rutland, VT: Tuttle, 1998.

Gutek, Gerald Lee. *Pestalozzi and Education.* Prospect Heights, IL: Waveland Press, 1999.

Haga, Tōru. *Kaiga no ryōbun: Kindai Nihon hikaku bunka-shi kenkyū.* Tokyo: Asahi Shimbun-sha, 1984.

Hall, Ivan Parker. *Mori Arinori.* Cambridge, MA: Harvard University Press, 1973.

Harada, Teishū. *Shōgaku tokuhon.* Tokyo: Bunkaido, 1888.

Hata, Takehiko. *Sokki! Jinsei ni wa yaku ni tattanai tokugi.* Tokyo: Kodansha, 2006.

Hawks, Francis. *Narrative of the Expedition of an American Squadron to the China Seas and Japan.* Washington, DC: A.O.P. Nicholson, 1856.

Hayashi, Shigeatsu. *Hayagakitori no shikata.* Tokyo: Ikebekatsuzō, 1885.

Headrick, Daniel. *When Information Came of Age: Technologies of Knowledge in the Age of Reason and Revolution, 1700–1850.* Oxford, UK: Oxford University Press, 2000.

Heidegger, Martin. *The Question Concerning Technology and Other Essays.* Translated by William Lovitt. New York: Harper & Row, 1977.

Hill, Christopher. "How to Write a Second Restoration: The Political Novel and Meiji Historiography." *Journal of Japanese Studies* 33, no. 2 (2007): 337–356.

Hirota, Shinobu. "Bijutsu toshite no shashin." *Waseda Bungaku* 15, no. 2 (1908).
Hisamatsu, Senichi, ed. *Meiji bungaku zenshū, vol 44: Ochiai Naobumi, Ueda Kazutoshi, Haga Yaichi, Fujioka Sakutarō shū.* Tokyo: Chikuma Shobō, 1968.
Hobsbawm, Eric. *The Age of Empire: 1875–1914.* New York, Vintage, 1989.
Hokenson, Jan Walsh. *Japan, France, and East-West Aesthetics: French Literature 1867–2000.* Madison, NJ: Fairleigh Dickinson University Press, 2004.
Holzmann, Gerald J., and Björn Pehrson. *The Early History of Data Networks.* Washington, DC: IEEE Computer Society Press, 1995.
Hosono, Masanobu. *Nagasaki Prints and Early Copperplates.* Translated by Lloyd Craighill. New York: Kodansha International, 1978.
Howland, Douglas. *Translating the West.* Honolulu: University of Hawai'i Press, 2002.
Hyōdō, Hiromi. *Koe no kokumin kokka Nihon.* Tokyo: NHK Books, 2000.

Inoue, Miyako. "Gender, Language and Modernity: Toward an Effective History of Japanese Women's Language." *American Anthropologist* 29, no. 3 (2002): 392–422.
———. *Vicarious Language: Gender and Linguistic Modernity in Japan.* Berkeley: University of California Press, 2006.
Inouye, Charles, ed. *Japanese Gothic Tales: Izumi Kyōka.* Honolulu: University of Hawai'i Press, 1996.
Inouye, Charles S. "Pictocentrism," *Yearbook of Comparative and General Literature* 40 (1992): 23–39.
Inwood, M.J., ed. *Hegel's Philosophy of Mind.* Translated by W. Wallace and A.V. Miller. Oxford: Clarendon Press, 2007.
Isawa, Shūji. *Shiwahō.* Tokyo: Dai-Nihon Tosho, 1901.
Itō, Kiyoshi, ed. *Sanyūtei Enchō to sono jidai (ten).* Tokyo: Waseda Daigaku Engeki Hakubutsukan, 2000.
Ivy, Marilyn. *Discourses of the Vanishing : Modernity, Phantasm, Japan.* Chicago: University of Chicago Press, 1995.
Iwanami Shoten Bungaku Henshūbu. *Bungaku zokan: Enchō no sekai.* Tokyo: Iwanami Shoten, 2000.

Jameson, Harold, ed. *Peonies Kana: Haiku by the Upasaka Shiki.* London: George Allen & Unwin Ltd., 1973.
Jay, Martin. *Downcast Eyes: The Denigration of Vision in Twentieth-Century French Thought.* Berkeley: University of California Press, 1994.
Johnston, Edward. *The Artistic Crafts Series of Technical Handbooks: Writing and Illuminating and Lettering.* New York: Macmillan, 1913.
Johnston, John, ed. *Literature, Media & Information Systems: Friedrich A. Kittler Essays.* Translated by William Stephen Davis. Amsterdam: G+B Arts, 1997.
Jones, H. J. *Live Machines: Hired Foreigners and Meiji Japan.* Vancouver: University of British Columbia Press, 1980.

Kamei, Hideo. *Meiji Bungaku-shi.* Tokyo: Iwanami Textbooks, 2000.
———. *Transformations of Sensibility.* Translated by Michael Bourdaghs. Ann Arbor: Michigan Monograph Series, 2001.

Kamiyama, Akira. "Customs of the Meiji Period and Kabuki's War Dramas." Translated by Joseph Ryan. *Comparative Theatre Review* 11, no. 1 (2012): 4–20.

Kanagawa-ken Kindai Bijutsukan, eds. *Kanagawa-ken bijutsu fudoki: bakumatsu Meiji shoki-hen*. Yokohama: Yurindō, 1970.

———. *Kanagawa-ken bijutsu fudoki: Takahashi Yū'ichi-hen*. Yokohama: Yurindō, 1972.

Kaneko, Kazuo. *Kindai nihon bijutsu kyōiku no kenkyū: Meiji, Taishōki*. Tokyo: Chūō Kōron, 2000.

———. "Meiji-ki zuga kyōiku ni okeru enpitsu-ga to mohitsu-ga." *Kokubungaku* 45, no. 8 (2000).

Kaneko Ryuichi, "Japanese Photography in the Early Twentieth Century." In *Modern Photography in Japan, 1915–1940*, edited by the Ansel Adams Adams Center. San Francisco: The Friends of Photography, 2001.

Karasawa, Tomitaro. *Kyōkasho no rekishi*. Tokyo: Sobunsha, 1981.

Karatani, Kojin. *The Origins of Modern Japanese Literature*. Translation supervised by Brett de Bary. Durham, NC: Duke University Press, 1993.

———. "Japan as Museum: Okakura Tenshin and Ernest Fenollosa." In *Japanese Art after 1945: Scream against the Sky*, edited by Alexandra Monroe. New York: Harry N. Abrams, 1994.

———. *The Structure of World History*. Translated by Michael Bourdaghs. Durham, NC: Duke University Press, 2014.

Katō, Shūichi. *A History of Japanese Literature: The Modern Years*. Translated by Don Sanderson. New York: Kodansha International, 1983.

Katō, Shūichi, and Maeda Ai. *Meiji media-kō*. Tokyo: Chūō Kōronsha, 1983.

Katō, Shūichi, and Maruyama Masao. *Honyaku no shisō*. Tokyo: Iwanami Shoten, 1991.

Kawamura, Minato. *Sakubun no naka no Dai-Nippon teikoku*. Tokyo: Iwanami Shoten, 2000.

Kawada Junzō, "'Hanashi' ga moji ni naru toki." *Kōtō denshōron*, vol. 2. Tokyo: Heibonsha, 2001.

Keene, Donald. *Dawn to the West: Japanese Literature in the Modern Era*, vols. 1–2. New York: Holt, Reinhart and Winston, 1984.

———, ed. *Modern Japanese Literature: From 1868 to the Present Day*. New York: Grove Press, 1956.

Kelly, John. "The 1847 Alphabet: An Episode of Phonotypy." In *Towards a History of Phonetics*, edited by R. E. Asher and E. J. A. Henderson, 248–264. Edinburgh, UK: Edinburgh University Press, 1981.

Kida, Junichirō. *Nihongo daihakubutsukan*. Tokyo: Just System, Inc., 1994.

———, ed. *Edogawa Rampo zuihitsu-sen*. Tokyo: Chikuma Bunko, 1994.

———. *Zusho Nihongo no kindai-shi: Gengo bunka no hikari to kage*. Tokyo: Just System, Inc., 1997.

Kinoshita, Naoyuki. *Iwanami kindai Nihon no bijutsu 4: Shashingaron*. Tokyo: Iwanami Shoten, 1996.

———. *Nihon no shashinka 1: Ueno Hikoma to bakumatsu no shashinka-tachi*. Tokyo: Iwanami Shoten, 1997.

Kitazawa, Noriaki. *Kyōkai no bijutsushi: "Bijutsu" keiseishi nōto*. Tokyo: Brucke, 2005.

Kittler, Friedrich. *Discourse Networks 1800/1900*. Translated by Michael Metteer with Chris Cullens. Stanford, CA: Stanford University Press, 1990.

———. *Gramophone, Film, Typewriter*. Translated by Geoffrey Winthrop-Young and Michael Wurtz. Stanford, CA: Stanford University Press, 1999.

———. *Optical Media: Berlin Lectures 1999*. Translated by Anthony Enns. Malden, MA: Polity Press, 2010.

Kittler, Friedrich, and Matthew Griffin. "The City Is a Medium," in *New Literary History* 27, no. 4, *Literature, Media and the Law* (Autumn 1996): 717–729.

Koike, Kazuo, et al. *Kanji mondai to mōji kōdo*. Tokyo: Ota Shuppan, 1999.

Kobayashi, Tadashi, ed. *Ukiyoe no rekishi*. Tokyo: Bijutsu Shuppansha, 1998.

Komagome, Takeshi. *Shokuminchi teikoku Nihon no bunka tōgō*. Tokyo: Iwanami Shoten, 1996.

Komatsu, Hideo. *Iroha uta*. Tokyo: Chūkō Shinsho, 1979.

Komori, Yōichi. *Nihongo no kindai*. Tokyo: Iwanami Shoten, 2000.

———. *Sōseki o yominaosu*. Tokyo: Chikuma Shinsho, 1994.

———. *Yuragi no Nihon bungaku*. Tokyo: NHK Books, 1998.

Kondō, Ichitarō, and Charles Terry, eds. *The Thirty-Six Views of Mount Fuji by Hokusai*. Tokyo: Heibonsha, 1968.

Kono, Kensuke. *Shomotsu no kindai: media no bungakushi*. Tokyo: Chikuma Shōbō, 1999.

Kōno, Shōichi, and Nakamura Mitsuo, eds. *Futabatei Shimei zenshū*. Tokyo: Iwanami Shoten, 1964–65.

Kornicki, Peter. *The Reform of Fiction in Meiji Japan*. London: Ithaca Press for the Board of the Faculty of Oriental Studies, Oxford University, 1982.

Kōsaka Masaaki, *Japanese Thought in the Meiji Era*. Translated by David Abosch. Tokyo: Pan-Pacific Press, 1958.

Krell, David Farrell, ed. *Martin Heidegger: Basic Writings*. San Francisco: Harper Collins, 1993.

Kunikida, Doppo. *The River Mist and Other Stories*. Translated by David Chibett. Kent, UK: P. Norbury, 1993.

Kunikida Doppo Zenshū Hensan Iinkai, eds. *Kunikida Doppo zenshū*. Tokyo: Gakushū Kenkyūsha, 1964–1969.

Lamarre, Thomas. "The Deformation of the Modern Spectator: Synaesthesia, Cinema and the Spectre of Race in Tanizaki," *Japan Forum* 11, no. 1 (1999): 23–42.

———. "Magic Lantern: Dark Precursor of Animation," *Animation* 6 (2011): 127–148.

———. *Uncovering Heian Japan: An Archeology of Sensation and Inscription*. Durham, NC: Duke University Press, 2000.

Lamarre, Thomas, and Kang Nei-hui, eds. *Impacts of Modernities (Traces 3)*. Hong Kong: Hong Kong University Press, 2004.

Laver, James. *French Painting and the Nineteenth Century*. London: B. T. Batsford Ltd., 1937.

Lessing, G. E. *Laocoon: An Essay on the Limits of Painting and Poetry*. Translated by E. C. Beasley. London: Brown, Green and Longmans, 1891.

Leutner, Robert. *Shikitei Samba and the Comic Tradition in Edo Fiction*. Cambridge, MA: Harvard University Press, 1985.

Levin, David Michael, ed. *Modernity and the Hegemony of Vision*. Berkeley: University of California Press, 1993.

Li, Takanori. *Hyōshō kūkan no kindai: Meiji "Nihon" no media hensei*. Tokyo: Shinyōsha, 2000.

Lounsbury, Thomas. *English Spelling and Spelling Reform*. New York: Harper & Brothers, 1909.

MacLachlan, Patricia. *The People's Post Office: The History and Politics of the Japanese Postal System, 1871–2010*. Cambridge, MA: Harvard University Asia Center, 2011.

Maeda, Ai. *Kindai dokusha no seiritsu*. Tokyo: Chikuma Shobō, 1989.

———. *Text and the City: Essays on Modernity*. Edited by James Fujii. Durham, NC: Duke University Press, 2004.

Maeda, Ai, and Yamada Yūsaku, eds. *Nihon bungaku taikei*, vol. 2: *Meiji seiji shōsetsu-shū*. Tokyo: Kadokawa Shoten, 1974.

Mair, Victor, ed. *Wandering on the Way: Early Taoist Tales and Parables of Chuang Tzu*. Honolulu: University of Hawaii Press, 1998.

Marra, Michael, ed. *Japanese Hermeneutics: Current Debates on Aesthetics and Interpretation*. Honolulu: University of Hawaii Press, 2002.

Marumaru Chinbun, eds. *Marumaru Chinbun*. Tokyo: Honpō shoseki, 1981–1985.

Maruyama, Heijirō. *Kotoba no shashinhō: Ichimei hikkigaku kaitei*. Osaka: Eigaku Jitaku Dokushūkai, 1885.

Masaoka, Chūsaburō, ed. *Shiki zenshū*. Tokyo: Kodansha, 1975.

Masaoka, Shiki. *Masaoka Shiki: Selected Poems*. Translated by Burton Watson. New York: Columbia University Press, 1997.

Matsui, Takako. *Shasei no henyō: Fontanesi kara Shiki, soshite Naoya e*. Tokyo: Meiji Shoin, 2002.

Mayeda Shinjirō. *Outlines of the History of Telegraphs in Japan*. Tokyo: Kokubunsha, 1892.

———. *A Short Sketch of the Progress of the Postal System*. Tokyo: Kokubunsha, 1892.

Mayr, Otto. *Authority, Liberty and Automatic Machinery in Early Modern Europe*. Baltimore: Johns Hopkins University, 1989.

McLuhan, Marshall. *The Gutenberg Galaxy*. Toronto: University of Toronto Press, 2000.

———. *Understanding Media*. Cambridge, MA: MIT Press, 2001.

Meech-Pekarik, Julia. *The World of the Meiji Print: Impressions of a New Civilization*. New York: Weatherhill, 1986.

Mertz, John Pierre. *Novel Japan*. Ann Arbor: University of Michigan Press, 2003.

Military Government Translation Center, trans. *Japanese Postal Laws*. New York: Naval School of Military Government and Administration, 194?.

Miller, J. Scott. *Adaptations of Western Literature in Meiji Japan*. New York: Palgrave, 2001.

Minamoto, Tsunanori. *Shinshiki sokkijutsu*. Tokyo: Aoki Suzandō, 1893.

Mitchell, Timothy. *Colonising Egypt*. Berkeley: University of California Press, 1991.

Miyago, Toshio. *Meiji no bijinga: E-hagaki ni miru Meiji no esupuri*. Kyoto: Kyoto Shoin Arts Collection, 1998.

Morris, Ivan. *World of the Shining Prince*. New York: Knopf, 1964.

Morris, Mark. "Buson and Shiki, Part I and Part II," *Harvard Journal of Asiatic Studies* 44, no. 2, 45, no. 1 (December 1984, June 1985): 381–425, 255–321.

Mosher, John. *Japanese Post Offices in China and Manchuria*. New York: Quarterman Publications, 1978.

Motoyama, Yukihiko. *Meiji kokka no kyōiku shishō*. Tokyo: Shibunkaku Shuppan, 1998.

Murakami, Yuji, dir. *Kokugo Gannen*. Japan: NHK, 1985. Television series.

Murakata, Akiko, ed. *The Ernest E. Fenollosa Papers: The Houghton Library, Harvard University, Japanese Edition*, vol 3. Tokyo: Museum Press, 1987.

Mutō, Teruaki. "Mori Arinori's 'Simplified English': A Socio-Historical Examination," *Forum of International Development Studies* 26 (2004):89–101.

Nagamine, Shigetoshi. *Dokusho kokumin no tanjō*. Tokyo: Nihon Editā Sukūru Shuppansha, 2003.

———. *Zasshi to dokusha no kindai*. Tokyo: Nihon Editor's School Shuppan-bu, 1997.

Nagata, Seiji, ed. *Fūkeiga 2: Hokusai bijutsukan*. Tokyo: Shūeisha, 1990.

———. *Hokusai manga*, vols. 1–3. Tokyo: Iwanani Shoten, 1987.

Naitō, Takashi. *Meiji no oto: Seiyōjin ga kiita kindai Nihon*. Tokyo: Chūō Kōron Shinsha, 2005.

Nakamura, Kikuji. *Kyōkasho no shakaishi: Meiji ishin kara haisen made*. Tokyo: Iwanami Shinsho, 1992.

Nakamura, Takafumi. *Shisen kara mita Nihon kindai*. Kyoto: Kyoto Daigaku Gakujutsu Shuppankai, 2000.

———. *"Shisen" kara mita Nihon kindai: Meiji-ki zuga kyōiku-shi kenkyū*. Tokyo: Tokyo Daigaku Gakujutsu Shuppan-kai, 2000.

Nakatani, Hajime. "Body and Signs in Early Medieval China." Ph.D. diss., University of Chicago, 2004.

Natsume, Sōseki. *And Then*. Translated by Norma Field. Baton Rouge: Louisiana State University Press, 1978.

———. *I Am a Cat*. Translated by Aiko Ito and Graeme Wilson. Boston: Tuttle Publishing, 2002.

———. *Natsume Sōseki zenshū*. Tokyo: Iwanami Shoten, 1994.

———. *Sanshiro*. Translated by Jay Rubin. Ann Arbor, MI: Center for Japan Studies, 2002.

Nietzsche, Friedrich. *Beyond Good and Evil*. London & Edinburgh: T. N. Foulis, 1914.

Nihon Bijutsukan Henshūbu, eds. *Nihon Bijutsukan: The Art Museum of Japan*. Tokyo: Shogakukan, 1997.

Nolletti, Arthur, Jr., and David Desser. *Reframing Japanese Cinema: Authorship, Genre, History*. Bloomington: Indiana University Press, 1992.

Nomura, Masaaki. *Rakugo no gengogaku*. Tokyo: Heibonsha, 2002.

Ochi, Haruo, ed. *Yano Ryūkei-shū*. Tokyo: Chikuma Shobō, 1970.

Oguma Eiji. *Tanitsu minzoku shinwa no kigen*. Tokyo: Shinchōsha, 1995.

Oka, Kazuo, and Tokieda Motoki. *Kokugo kokubungaku shiryō zukai daijiten*. Tokyo: Zenkoku Kyōiku Zusho Kabushigaisha, 1968.

Okrent, Arika. *In the Land of Invented Languages*. New York: Spiegel and Grau, 2010.

Ōkubo, Toshiaki, ed., *Mori Arinori zenshū*, vol. 1. Tokyo : Senbundō Shoten, 1972.

———, ed. *Nishi Amane zenshū*, vol. 4. Tokyo: Munetaka Shobō, 1986.

Ōkuma, Shigenobu, ed. *Fifty Years of New Japan*, vols. 1–2. Translated by Marcus Huish. London: Smith, Elder & Co., 1910.

O'Malley, Michael. *Keeping Watch: A History of American Time*. New York: Viking, 1990.

Ortolani, Benito. *Japanese Theatre from Shamanistic Ritual to Contemporary Pluralism*. Princeton, NJ: Princeton University Press, 1995.

Osa, Shizue. *Kindai Nihon to kokugo nashionaruizumu*. Tokyo: Yoshikawa Kōbunkan, 1998.

Ōsawa, Shigeo. "Mori Arinori no chōshō—*shōsetsuka no kigen* hoi 'hihan,'"*Juryoku*, 1 (2002).

Perlman, David. "Physicists Convert First Known Sound Recording." *San Francisco Chronicle*, March 29, 2008.

Peters, John Durham. *Speaking into the Air: A History of the Idea of Communication*. Chicago: University of Chicago Press, 1999.

Piovesana, Gino. *Recent Japanese Philosophical Thought, 1862–1996*. Avon, UK: Japan Library, 1997.

Pitman, Isaac. *A Manual of Phonography, or, Writing by Sound: A Natural Method by Signs that Represent Spoken Sounds*. London: F. Pitman, 1883.

Ravina, Mark. *Land and Lordship in Early Modern Japan*. Stanford, CA: Stanford University Press, 1999.

———. "Japanese State Making in Global Context." In *State Making in Asia*, edited by Robert Boyd, 35–52. New York: Routledge, 2006.

Reed, Thomas Allen. *The Shorthand Writer*. London: Isaac Pitman & Sons, 1892.

Richter, Giles. "Marketing the Word: Publishing Entrepreneurs in Meiji Japan, 1870–1912." Ph.D. diss., Columbia University, 1999.

Rimer, J. Thomas, ed. *Mori Ōgai: Youth and Other Stories*. Honolulu: University of Hawaii Press, 1994.

Rodd, Laurel, and Mary Catherine Henkenius, trans. *Kokinshū: A Collection of Poems Ancient and Modern*. Princeton: Princeton University Press, 1984.

Rubin, Jay. *Injurious to Public Morals*. Seattle: University of Washington Press, 1984.

Ryan, Marleigh Grayer. *Japan's First Modern Novel*. Ann Arbor, MI: Center for Japan Studies, 1990.

Sakai, Naoki. *Kakko no koe: Jyūhasseiki Nihon no gensetsu ni okeru gengo no chii*. Translated by Kawata Jun et al. Tokyo: Ibunsha, 2003.

———. *Shisan sareru Nihongo Nihonjin*. Tokyo: Shinyōsha, 1996.

———. *Translation and Subjectivity*. Minneapolis: University of Minnesota Press, 1997.

———. *Voices of the Past*. Ithaca, NY: Cornell University Press, 1991.

Sakurai, Susumu. *Edo no noizu: kangoku toshi no hikari to yami*. Tokyo: NHK Books, 2000.

Satō, Dōshin. *Meiji kokka to kindai bijutsu*. Tokyo: Yoshikawa Hirofumi, 1999.

———. *Modern Japanese Art and the Meiji State: The Politics of Beauty*. Translated by Hiroshi Nara. Los Angeles: Getty Research Institute, 2011.

Saussure, Ferdinand. *Course in General Linguistics*. Translated by Roy Harris. New York: Open Court Classics, 1998.

Sayers, Robert. "Ainu: Spirit of a Northern People," *American Anthropologist* 102, no. 4 (2000): 877–882.

Schivelbusch, Wolfgang. *The Railroad Journey: The Industrialization of Time and Space in the Nineteenth Century*. Berkeley: University of California Press, 1986.

Schmid, Andre. *Korea between Empires, 1895–1919*. New York: Columbia University Press, 2002.

Sconce, Jeffrey. *Haunted Media: Electronic Presence from Telegraphy to Television*. Durham, NC: Duke University Press, 2000.

Screech, Timon. *The Lens within the Heart: The Western Scientific Gaze and Popular Imagery in Later Edo Japan*. New York: Cambridge University Press, 1996.

———. *The Shogun's Painted Culture*. London: Reaktion Books, 2000.

Sedgwick, Eve. *Between Men: English Literature and Male Homosocial Desire*. New York: Columbia University Press, 1985.

Seeley, Christopher. *A History of Writing in Japan*. Honolulu: University of Hawai'i Press, 2000.

Seltzer, Mark. *Bodies and Machines*. New York: Routledge, 1992.

———. *Serial Killers*. New York: Routledge, 1998.

Shaw, George Bernard. *Pygmalion*. New York: Dover Thrift Editions, 1994.

Shi, Gang. *Shokuminchi shihai to Nihongo*. Tokyo: Sangensha, 1993.

Shinano Kyōikukai, eds. *Isawa Shūji senshū*. Nagano: Shinano Kyōikukai, 1958.

Shinohara, Hiroshi. *Meiji no yūbin tetsudō basha*. Tokyo: Yūshōdō Shuppan, 1987.

Shōyō Kyōkai, eds. *Shōyō senshū*. Tokyo: Dai-ichi Shobo, 1979.

Siegert, Bernhard. *Relays: Literature as an Epoch of the Postal System*. Translated by Kevin Repp. Stanford, CA: Stanford University Press, 1999.

Shimane Kenritsu Daigaku Nishi Amane kenkyū-kai. *Nishi Amane to Nihon no kindai*. Tokyo: Perikansha, 2005.

Smith, Henry Dewitt. *Kiyochika: Artist of Meiji Japan*. Santa Barbara: Santa Barbara Museum of Art, 1988.

Sōgō, Masaaki, and Hida Yoshifumi, eds. *Meiji no kotoba jiten*. Tokyo: Tokyodō, 1986.

Spencer, Herbert. *Philosophy of Style*. New York: Allyn and Bacon, 1892.

Standage, Tom. *The Victorian Internet*. New York: Berkley Books, 1998.

Sterne, Laurence. *A Sentimental Journey*. New York: Penguin Classics, 2002.

Sturge, E.A. "Language Study," *Taiyō* 10: 2 (July 1904): 6–10 (English section).

Suga, Hidemi. *Nihon kindai bungaku no "tanjō": genbun itchi, undō to nashionarizumu*. Tokyo: Ota Shuppan, 1995.

Susskind, Flora. *Cinematograph of Words: Literature, Technique, and Modernization in Brazil*. Stanford, CA: Stanford University Press, 1997.

Suzuki, Yukizō, ed. *Sanyūtei Enchō zenshū*. Tokyo: Sekai Bunko, 1963.

Sweet, Henry. *A Primer of Spoken English*. Oxford: Clarendon Press, 1890.

Tai, Eika. "Kokugo and Colonial Education in Taiwan," *Positions: East Asia Cultures Critique* 7, no. 2 (1999): 503–540.

Takahashi, Seori. *"Byōsha, suketchi suru seishin,"* *Kokubungaku* 5, no. 8 (July 2000).

Takahashi Yasuo. *Media no akebono: Meiji kaikoku-ki no shuppan, shinbun monogatari.* Toyko: Nihon Keizai Shimbunsha, 1994.

Takeda Akira. *Chūgoku shōsetsu-shi nyūmon.* Tokyo: Iwanami Textbooks, 2002.

Tanaka, Atsushi. "Shasei ryokō." In *Nihon Bijutsukan: The Art Museum of Japan,* edited by Nihon Bijutsukan Henshūbu. Tokyo: Shogakukan, 1997.

Tanaka, Stefan. *New Times in Modern Japan.* Princeton: Princeton University Press, 2006.

Tanakadate, Aikitsu. *Metoru-hō no rekishi to genzai no mondai.* Tokyo: Iwanami Shoten, 1934.

Tanizaki, Junichirō. *Bunshō tokuhon.* Tokyo: Chūkō Bunko, 1991.

Tayama, Katai. *The Quilt.* Translated by Kenneth Henshall. Tokyo: University of Tokyo Press, 1981.

———. *Tayama Katai zenshū.* Tokyo: Bunseidō Shoten, 1974.

Terada, Tōru. "Kindai bungaku to Nihongo," republished in *Iwanami kōza Nihon bungaku-shi.* Tokyo: Iwanami Shoten, 1958–1959.

Tschudin, Jean-Jacques. "Danjurō's *katsureki-geki* and the Meiji 'Theatre Reform' Movement," *Japan Forum,* 11, no. 1 (1999): 83–94.

Thoreau, Henry David. *Walden.* New York: Oxford University Press, 1997.

To, Wing-kai. "Bridgewater Normal School and Isawa Shūji's Reforms of Modern Teacher Education in Meiji Japan," *Higashi Ajia bunka kōshō kenkyū* 2 (2009): 413–421.

Tomasi, Massimiliano. *Rhetoric in Modern Japan: Western Influences on the Development of Narrative and Oratorical Style.* Honolulu: University of Hawai'i Press, 2004.

Toyama, Usaburō. *Nihon yōga-shi: dai-ichi maki, Meiji zenki.* Tokyo: Nichieki Shuppansha, 1978.

Tsubouchi, Shōyō. "Bi to wa nan zo ya," *Gakugei Zasshi* 4 (1886): 219

———. *Essence of the Novel.* Translated by Nanette Twine. Brisbane: University of Queensland, 1981.

Tsubouchi Yūzō, ed. *Meiji no bungaku, dai-10 maku: Yamada Bimyō.* Tokyo: Chikuma Shobō, 2001.

Tsurumi, E. Patricia. *Japanese Colonial Education in Taiwan.* Cambridge, MA: Harvard University Press, 1977.

Tucker, Anne Wilkes, et al., eds. *The History of Japanese Photography.* New Haven, CT: Yale University Press, 2003.

Twain, Mark. *What Is Man?* New York: Oxford University Press, 1996.

Twine, Nanette. *Language and the Modern State: The Reform of Written Japanese.* New York: Routledge, 1991.

Ueda, Atsuko. *Concealment of Politics, Politics of Concealment: The Production of "Literature" in Meiji Japan.* Stanford, CA: Stanford University Press, 2007.

———. "The Production of Literature and the Effaced Realm of the Political," *Journal of Japanese Studies* 31, no. 1 (2005): 61–88.

Unger, J. Marshall. *Ideogram: Chinese Characters and the Myth of Disembodied Meaning.* Honolulu: University of Hawai'i Press, 2004.

Vincent, Keith. "Writing Sexuality: Heteronormativity, Homophobia, and the Homosocial Subject in Modern Japan." Ph.D. diss., Columbia University, 2000.

Watson, Burton. *Masaoka Shiki: Selected Poems*. New York: Columbia University Press, 1998.

Webster, Noah. *The American Spelling Book*. New York: Teachers College, Columbia University, 1962.

———. *Dissertations on the English Language*. Gainesville: Scholars' Facsimiles & Reprints, 1951.

Williams, Raymond. *Country and City*. London: Oxford University Press, 1975.

Woolf, Virginia. *Selected Essays*. New York: Oxford University Press, 2008.

Yamaguchi, Osamu. *Maejima Hisoka*. Tokyo: Yoshikawa Hirofumikan, 1990.

Yamaguchi, Seiichi, ed. *Fenollosa bijutsu ronshū*. Tokyo: Chūō Kōron Bijutsu Shuppansha, 1989.

Yamamoto, Masahide. *Genbun itchi no rekishi ronkō*. Tokyo: Ofusha, 1971.

———. *Kindai buntai hassei no shiteki kenkyū*. Tokyo: Iwanami Shoten, 1962.

———. *Kindai buntai keisei shiryō shūsei*. Tokyo: Ofusha, 1978.

Yanagita, Kunio. *Tōno monogatari*. Tokyo: Kyōdo Kenkyūsha, 1935.

Yano, Ryūkei. *Nihon buntai moji shinron*. Tokyo-fu: Hōchisha, 1886.

———. *Yano Ryūkei shiryōshū*, vol. 1. Oita: Oita-ken Kyōiku Iinkai, 1996.

———. *Keikoku Bidan*, vols. 1–2. Tokyo: Hōchi Shimbunsha, 1883–1884.

Yasuda Toshiaki. *Kindai Nihongo gengo kenkyū saikō*. Tokyo: Sangensha, 2000.

———. *Kokugo no kindaishi: teikoku Nihon to kokugo gakushatachi*. Tokyo: Chūō Kōron Shinsha, 2006.

———. *Shokuminchi no naka no "kokugogaku."* Tokyo: Sangensha, 1997.

Yasukuni Jinja, eds. *Yasukuni daihyakka: watashitachi no Yasukuni jinja*. Tokyo: Office of Yasukuni Shrine, 1992.

Yūseishō, eds. *Yūsei hyakunenshi shiryū*, vols. 19–26. Tokyo: Yoshikawakōbunkan, 1969–1971.

Index

Italic page numbers refer to figures.

Harvard East Asian Monographs
(most recent titles)

Harvard East Asian Monographs

Harvard East Asian Monographs

Harvard East Asian Monographs

Harvard East Asian Monographs

Harvard East Asian Monographs

Harvard East Asian Monographs